SpringerBriefs in Applied Sciences and Technology

More information about this series at http://www.springer.com/series/8884

Muhammad Azhar Ali Khan
Anwar Khalil Sheikh · Bilal Suleiman Al-Shaer

Evolution of Metal Casting Technologies

A Historical Perspective

 Springer

Muhammad Azhar Ali Khan
Mechanical Engineering Department
King Fahd University of Petroleum
    and Minerals
Dhahran
Saudi Arabia

Anwar Khalil Sheikh
Mechanical Engineering Department
King Fahd University of Petroleum
    and Minerals
Dhahran
Saudi Arabia

Bilal Suleiman Al-Shaer
Axles, Foundries and Spare Parts Factory
    MASABIK
Dammam
Saudi Arabia

ISSN 2191-530X   ISSN 2191-5318 (electronic)
SpringerBriefs in Applied Sciences and Technology
ISBN 978-3-319-46632-3   ISBN 978-3-319-46633-0 (eBook)
DOI 10.1007/978-3-319-46633-0

Library of Congress Control Number: 2016955073

Printed on acid-free paper

This Springer imprint is published by Springer Nature
The registered company is Springer International Publishing AG
The registered company address is: Gewerbestrasse 11, 6330 Cham, Switzerland

Contents

Evolution of Metal Casting Technologies—A Historical Perspective

1 Introduction

Metalcasting has been serving the manufacturing industry for years. It is used for manufacturing a range of products from being simple to exceptionally intricate. The challenges associated with the process have led to many advancements in metal casting technologies over time. During this period, different cast materials and casting processes have been evolved making it possible to design and manufacture almost any product with high quality. It is important to analyze the progress of metalcasting in past, trends in the field at present, and envisage the future of these technologies for continuous improvement. This book provides a comprehensive knowledge of metalcasting technologies by reviewing its historical evolution at first. A number of processes have emerged using innovative techniques for handling molten metals making it crucial to select the appropriate process for uninterruptable production. Therefore, a detailed description of metalcasting processes under different categories is presented next. A thorough discussion on foundry practices such as mold materials and molding techniques, pattern making and cores, cast alloys, furnaces, pouring, cleaning and heat treatment operations etc. is also included. Since, a host of process parameters are involved, castings are often produced with defects which needs to be minimized or eliminated. This issue is addressed here by explaining various types of casting defects along with their possible causes and remedies. As a final point, key considerations in casting design are elaborated. The underlying idea behind this study is to let the readers ingrain the fundamental knowledge of metalcasting technologies in all above mentioned areas so that the integrity and quality of castings can be enhanced.

© The Author(s) 2017
M.A.A. Khan et al., *Evolution of Metal Casting Technologies*,
SpringerBriefs in Applied Sciences and Technology,
DOI 10.1007/978-3-319-46633-0_1

2 Historical Evolution of Metalcasting

Metalcasting is one of the primitive manufacturing processes which was developed based on fire-using technologies (or pyrotechnologies). Initially, pyrotechnologies have been used to improve workability of stone, make plaster by burning lime, and produce ceramics by firing the clay. Metalworking, however, started 10,000 years ago when the earliest metal objects are found to be wrought rather than cast [1, 10]. Evidences of such metal objects are found in the form of decorative pendants and beads which were formed by hammering native copper. Table 1 summarizes the advancements in metalworking over these 10,000 years. Metal casting, on the other hand, dates back to 5000 and 3000 B.C. which refers to Chalcolithic period during which metals were melt for castings together with the experimentation of smelting copper. Initial molds were made from smooth textured stones as shown in Fig. 1, resulting in fine cast products which could be witnessed in the museums and archeological exhibitions. Both single and multifaceted (carved on both sides of a rectangular piece of stone) molds were developed from stones to produce castings that are not necessarily flat. However, multifaceted molds were more popular for being portable and economical in terms of utilizing a suitable piece of stone [1].

Followed by Chalcolithic period, the Bronze Age started in Near East around 3000 B.C. during which alloying elements were added to copper. The first bronze alloy reported was a mix of copper (Cu) with 4–12 %percent of arsenic (As), thus forming a silvery appearance of the cast surface as a result of inverse segregation of the arsenic-rich low-melting phase to the surface. Next, 5–10 % of tin (Sn) was added to copper which was advantageous to lower the melting point, improve strength, deoxidize the melt, and produce a fine and easily polished cast surface capable of reproducing the features of the mold with exceptional fidelity, often desirable for art castings. The Bronze Age lasted for 1500 years during which mankind had its first exposure to elemental ores such as tin, copper and silver. Also, lost wax castings of small parts of bronze and silver is reported during this age between 3000 and 2500 B.C. in the Near East region. The earliest known casting in existence is a copper frog as shown in Fig. 2, probably cast in 3200 B.C. in Mesopotamia region [23]. The intricate design and three-dimensional characteristics suggests that it was created using sand casting process instead of using a permanent stone mold [10]. The beginning of Bronze Age in Far East was about 2000 B.C., a millennium after its emergence in the Near East region. Casting was main forming process in China with little evidences of other metalworking operations before 500 B.C. The complexity of antique cast bronze ritual vessels suggested that they must have been produced using lost wax casting method.

The use of cast iron in Chinese statuary began in 600 B.C. High phosphorus and sulfur contents in cast iron resulted in melting temperature comparable to bronze together with unusual fluid behavior in the molten state [1]. In Europe, cast iron was introduced as a casting alloy between 1200 and 1450 A.D. Between 15th and 18th century, some other important developments in the casting industry include but not limited to introduction of sand as mold material in 16th century, production of cast

Table 1 Development in use of materials and metalcasting [1, 11]

Date	Development	Location
9000 B.C.	Earliest metal objects of wrought native copper	Near East
6500 B.C.	Earliest life-size statues, of plaster	Jordan
5000–3000 B.C.	Chalcolithic period: melting of copper; experimentation with smelting	Near East
3000–1500 B.C.	Bronze age: arsenical copper and tin bronze alloys	Near East
3000–2500 B.C	Lost wax casting of small objects	Near East
2000 B.C.	Bronze age	Far East
600 B.C.	Cast iron	China
1200–1450 A.D.	Introduction of cast iron	Europe
16th century	Sand introduced as mold material	France
1709 A.D.	Cast iron produced with coke as fuel, Coalbrookdale	England
1735 A.D.	Great bell of the Kremlin cast	Russia
1740 A.D.	Cast steel developed by Benjamin Huntsman	England
1779 A.D.	Cast iron used as architectural material, Ironbridge Gorge	England
1809 A.D.	Development of centrifugal casting	England
1818 A.D.	Production of cast steel by crucible process	U.S.
1837 A.D.	Development of first molding machine	U.S.
1849 A.D.	Development of a die casting machine	–
1863 A.D.	Metallography of casting surfaces	England
1870 A.D.	Development of sand blasting for large castings	U.S.
1876 A.D.	Aluminum started to use as cast material	U.S.
1887 A.D.	Development of oven for core drying	–
1898 A.D.	Mold development with bonded sand	England
1899 A.D.	Electric Arc Furnace for commercial production of castings	France
1900 A.D.	Low pressure permanent mold casting	England
1907 A.D.	Heat treatment and artificial aging to improve cast aluminum alloys	Germany
1910 A.D.	Matchplates and Jolt Squeezing machines	–
1925 A.D.	X-ray radiography for casting quality control	U.S.
1940 A.D.	Chvorinov's Rule	–
1944 A.D.	First heat-reactive, chemically-cured binder	Germany
1948 A.D.	Ductile Iron castings in industrial applications	U.S.
1950s	High pressure molding	–
Mid 1950s	Squeeze casting process	Russia
1960 A.D.	Development of Furan as core binder	–
1968 A.D.	Cold box process for mass production of cores	–
1971 A.D.	Vacuum forming or the V-process method	Japan
1974 A.D.	In-mold process for ductile iron treatment by Fiat	U.S.

(continued)

Table 1 (continued)

Date	Development	Location
1980s	Development and commercialization of a solidification software	–
1988 A.D.	Rapid prototyping and CAD/CAM technologies	–
1993 A.D.	Plasma ladle refining (melting and refining in one vessel)	U.S.
Mid 1990s	Microstructure simulations	–
1996 A.D.	Cast Metal Matrix Composites in automobile applications	England
End 1990s	Stress and distortion simulation of castings	–
2001 A.D.	Software development by NASA and Department of Energy/OIT	U.S.
2006 A.D.	Casting simulations coupled to mechanical performance simulations	–

Fig. 1 Mold prepared by smooth textured stone with an axe [1]

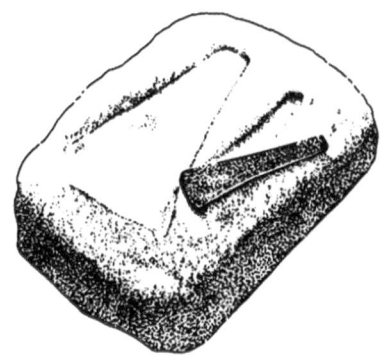

Fig. 2 The earliest known casting in existence "A Copper Frog", cast in 3200 B.C. [23]

(a)   (b)

Fig. 3 The Iron Bridge (**a**) across the Severn River at Ironbridge Gorge (**b**) Detail of the Iron Bridge showing the date, 1779 [1]

iron with coke as fuel at Coalbrookdale (1709 A.D.), casting of the Great Bell of the Kremlin (1735 A.D.), and the development of cast steel by Benjamin Huntsman (1740). The cast iron produced at Coalbrookdale was first used as an important structural material in building the famous iron bridge as shown in Fig. 3 and in other architectural applications during that period.

Although the development until 18th century was significant, cast materials and casting technologies progressed substantially from 19th century onwards. A new method of centrifugal casting was developed in 1809 A.D. in England. In 1818 A.D. first U.S. cast steel was produced by the crucible process at Valley Forge Foundry. The extraction of aluminum from aluminum chloride is also reported during the same century. The first molding machine was developed by S. Jarvis Adams Co. in Pittsburgh and was available in markets by 1837 A.D. The urge to further develop the casting process led to the development of a die casting machine to supply rapidly cast lead type for newspaper in 1849 A.D. Examination of castings was started by the development of metallography in 1863 A.D. which enabled foundrymen to polish, etch and physically analyze the castings through optical microscopy. Next, sand blasting was developed for large castings in 1870 A.D. By 1876 A.D. aluminum was started to use as cast material and its first architectural application was reported in 1884 A.D. when a cast aluminum pyramid was mounted on the tip of Washington Monument. Some other developments during the 19th century are oven for core drying (1887 A.D.), non-art application of lost wax casting method to produce dental inlays (1897 A.D.), mold with bonded sand for salt-water piping system (1898 A.D.), and electric arc furnace for commercial production of castings (1899 A.D.).

In early 1900s, first patent for low pressure permanent mold casting was issued in England. American Foundry Association (AFS) produced rail wheels by centrifugal casting process for the first time. The advancement in the field continued with a die casting machine patented by H.H. Doehler patents in 1905. Heat treatment and artificial aging was proposed in 1907 to improve the properties of cast aluminum alloys. In 1910, jolt squeezing machining became possible through the development of matchplates. Another important development in 1915 was the experimentation on bentonite clay due to its exceptional high green and dry strength. In addition to that furnaces for non-ferrous melting were also developed during 1910s. The quality of casting was first examined through X-ray radiography in 1925 after which all military aircraft castings had to pass X-ray inspection for acceptance by 1940. The development of mathematical relationships between casting geometry and solidification time was established by Chvorinov in 1940. Also, statistical process control was started to use for casting quality control and assurance in 1940s in the United States. The research on binders resulted in first heat-reactive, chemically-cured binder in Germany in 1944 for rapid production of mortar and artillery shells during World War II. By 1948, ductile iron was not just limited to laboratory castings and started to use as a cast material in industrial applications.

In order to increase the mold hardness (density), high pressure molding was experimented in 1950s. Hotbox system to prepare and cure the cores simultaneously was introduced in 1953 thereby eliminating the need of dielectric drying ovens. In mid 1950s, squeeze casting process originated from Russia. In addition to that a full mold process was developed in 1958, known as lost foam casting, where the pattern and gating systems were carved from expanded polystyrene (EPS) and placed into a green sand mold. During 1960s, Furan was developed as a binder to be used in core production. Also shell flake resin was introduced in 1963 and it eliminated the need for different solvents. In 1968, a new method called "cold box process" was developed for mass production of cores in foundries. The late 20th century i.e. 1970–1999 brought more advancements to metal casting such as development of vacuum forming or the V-process method in 1971 to produce molds using unbonded sand by using vacuum. In 1974, Fiat developed an in-mold process for ductile iron treatment. During 1980s, it was started to investigate the casting processes computationally an example of which is the development and commercialization of a solidification software. In late 1980s, casting solidification software gained acceptance in foundries resulting in optimization of quality and cost of casting process in virtual reality. An important development during this period was 3D visualization techniques followed by the rapid prototyping and CAD/CAM technologies in 1988 which significantly reduced the time of tool development.

Plasma ladle refining (melting and refining in one vessel) was introduced for the first time in 1993 at Maynard Steel Casting Company in Milwaukee, WI. In order to cast large components through lost foam castings, a low-expansion synthetic mullite sand is patented by Brunswick Corp. in 1994. Microstructure simulations in mid 1990s enabled designers to analyze effects of metallurgy and predict and control mechanical properties of cast components. Cast metal matrix composites (MMCs) were used in automobile applications such as brake rotors for the first time in 1996. In the same year, General Motors Corp. developed a non-toxic and recyclable, water-soluble biopolymer-based core sand binder. Casting simulation developed further towards the end of 1990s by stress and distortion simulation. As a result, generation and distribution of residual stresses in the cast component could be well understood which allowed to control casting distortion, reduce residual stresses, minimize defects such as hot tears and cracking, minimize mold distortion and improve mold life. In 2001, a physics-based software was developed by NASA and Department of Energy/OIT capable of predicting the filling of EPS patterns and sand cores when process variable are changed. In 2006, a new approach was developed which emphasized on developing an accept/reject criteria for castings by integrating casting simulation with mechanical performance simulations. Integrated simulations are currently being researched with an aim to improve integrity and quality, which eventually result in reliable operation of cast parts in service.

3   Casting Methods and Processes

Metal casting process is considered to be one of the simplest and direct method of producing a near net shape product. The process essentially needs a mold cavity (made up of sand, ceramic or even steel) of the desired shape where molten metal is poured to get the cast product. In its most usual form, the molten metal is supplied to the mold cavity through pouring basin followed by running system which allows adequate flow of molten metal within the cavity. During solidification, most metals experience shrinkage and the additional amount of molten metal is supplied through risers or feeders. These additional features are added to casting system in order to produce sound (pore free) products with minimum defects (sand inclusions, slag, cracks, etc.). Figure 4 shows the schematic diagram of a typical metal casting process.

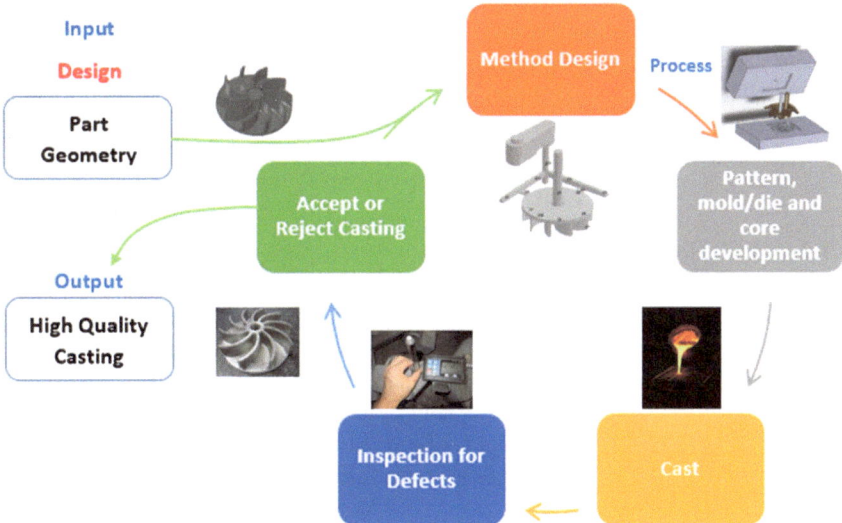

Fig. 4 Schematic diagram of a metalcasting process

Modern metal casting is classified based on a variety of parameters. In fact the casting processes can be distinguished based on (I) the type of mold such as sand, permanent, etc., (II) the flow of molten metal in mold cavity under the action of gravity, vacuum, pressure, (III) state of the metal i.e. fraction of metal which is liquid, (IV) the state of the mold cavity itself such as solid, gas, air or vacuum. In general, two or more of these processes are combined together. For example, use of green-sand molds and chemically bonded sand molds, semi-permanent molds where mold contains sand and metal components, high pressure or low pressure die casting etc. Figure 5 shows a detailed hierarchical classification of various casting processes. Table 2 summarizes some of the casting processes together with mold and metal details. A detailed description of different casting processes is provided in the forthcoming sub-sections.

3.1 Expendable Mold, Permanent Pattern Casting Processes

3.1.1 Sand Casting

Sand casting and sand molds are quite popular in metal casting industry owing to numerous advantages such as low capital investments, casting of fairly complex parts, production of large components as well as small batches etc. [6, 8]. Sand casting process begins with packing moist sand around a pattern, which is then removed to create the mold. Next, molten metal is poured into the mold cavity. The necessary molten metal during solidification is provided by risers. The mold is then broken to remove the cast part. Most of the ferrous metals and aluminum alloys are cast using sand casting, whereas difficulties may encounter in casting lead, tin and zinc alloys, beryllium, titanium and zirconia alloys and others [8]. Typical products of sand casting are engine blocks, machine tool bases, manifolds, pump housings and cylinder heads etc. Figure 6 depicts a typical sand casting operation together with all components in a mold cavity such as cope, drag, parting line, riser, sprue, pouring basin, etc.

3.1.2 Shell Molding

Shell molding process uses a heated metal pattern which is placed over a box of thermosetting resin-coated sand. Curing of sand is done by inverting the box for a fixed time. The excess sand is removed by inverting the box again. Next, the shell is removed from the pattern and joined with already made other half. They are supported in a flask by an inert material ready for casting. Shell molding technique can be used to cast most of the metals except: lead, zinc, magnesium and titanium

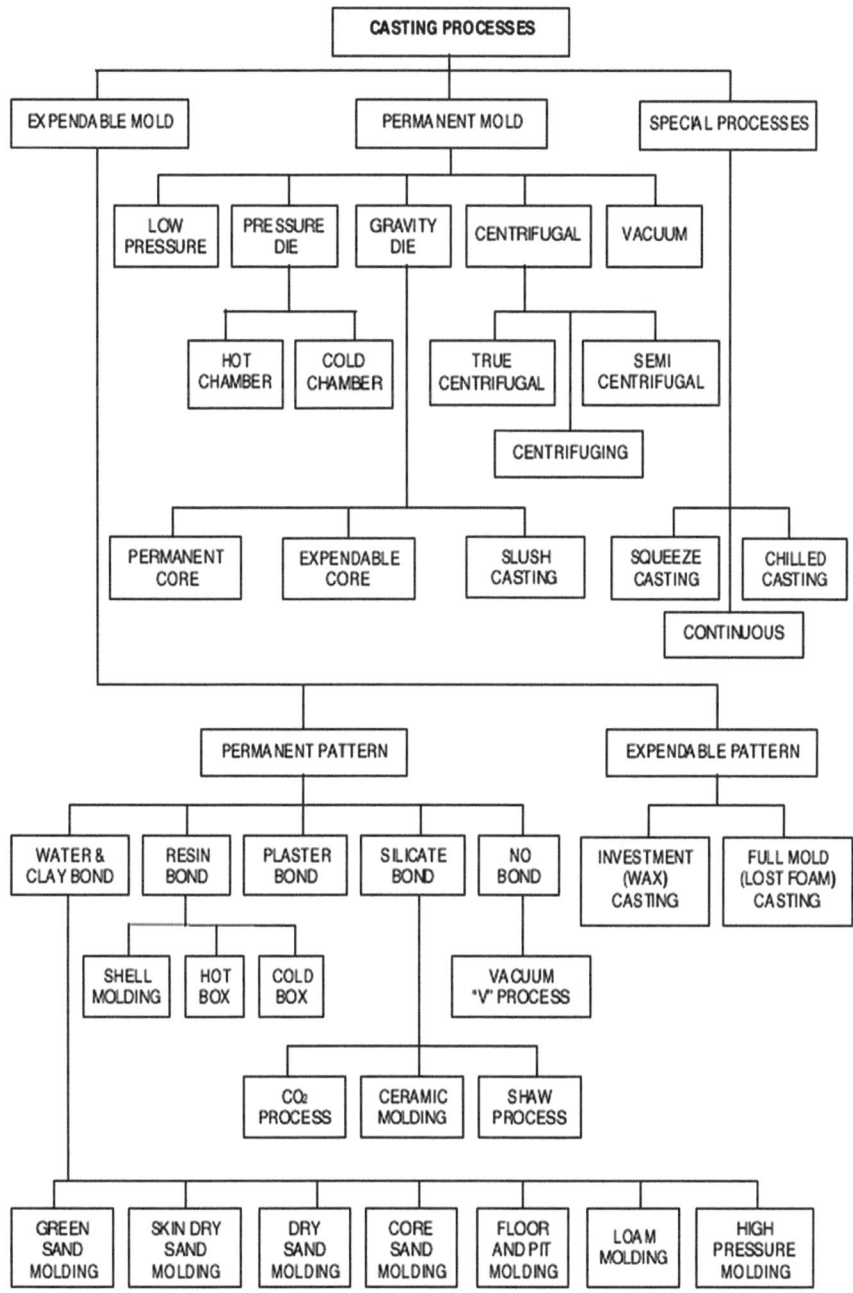

Fig. 5 Hierarchical classification of various casting processes [26]

Table 2 Casting processes, molds and metal details [18]

Casting process	(a) Mold material	(b) Manner of entry	(c) State of metal	(d) State of mold cavity
Sand casting	Sand (bonded with clay and water or chemicals)	Gravity	100 % Liquid	Air
Permanent mold	Metal	Gravity	100 % Liquid	Air
Die casting	Metal	Pressure	100 % Liquid	Air
Investment	Ceramic	Gravity	100 % Liquid	Air, Vacuum, Gas
Lost foam EPC	Sand (unbonded)	Gravity	100 % Liquid	Styrofoam, PMMA
Thixocasting Rheocasting	Metal	Pressure	>50 % Liquid Balance Solid	Air
Cosworth	Sand	Vacuum	100 % Liquid	Air
V process	Sand (unbonded with vacuum and enclosing plastic film)	Gravity	100 % Liquid	Air
Centrifugal	Metal, graphite	Centrifugal forces	100 % Liquid	Air, Gas Shroud
Ingot—NOT cast to Shape	Metal or electro-magnetic Field	Gravity	100 % Liquid	Air or Gas Shroud

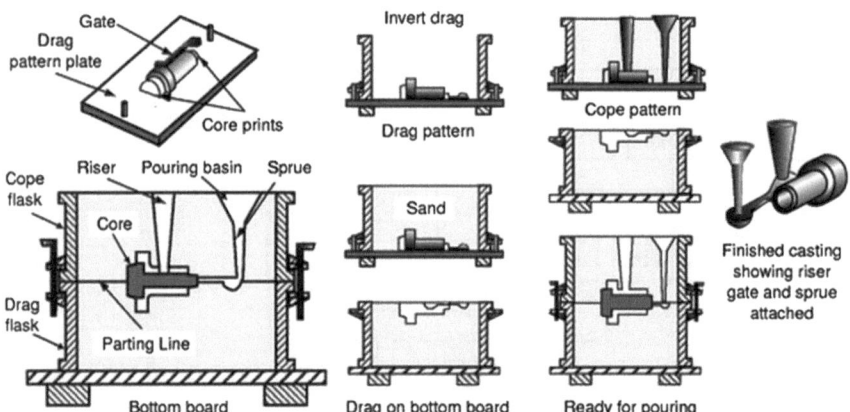

Fig. 6 Sand casting process [8]

alloys. Some of the castings produced by this technique are small mechanical parts requiring high precision, gear housings, cylinder heads, connecting rods transmission components etc. Figure 7 shows the process details of shell molding technique together with the finished cast product.

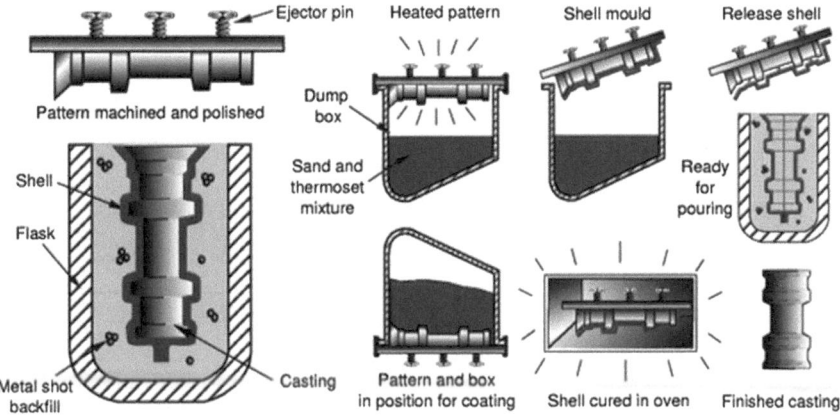

Fig. 7 Shell moulding process [8]

3.1.3   Plaster Mold Casting

Plaster mold casting utilizes a precise metal pattern which generates a two-part mold, commonly made up of gypsum slurry material. Moisture in the mold is removed by baking it in the oven. Molten metal is then poured into the plaster mold and allow to cool and solidify. Final cast part is removed by breaking the mold. This casting technique is limited to metals having low melting temperature owing to degradation of plaster mold at elevated temperatures. Some typical applications are gear blanks, valve parts, wave guide components, etc. The details of process are shown in Fig. 8.

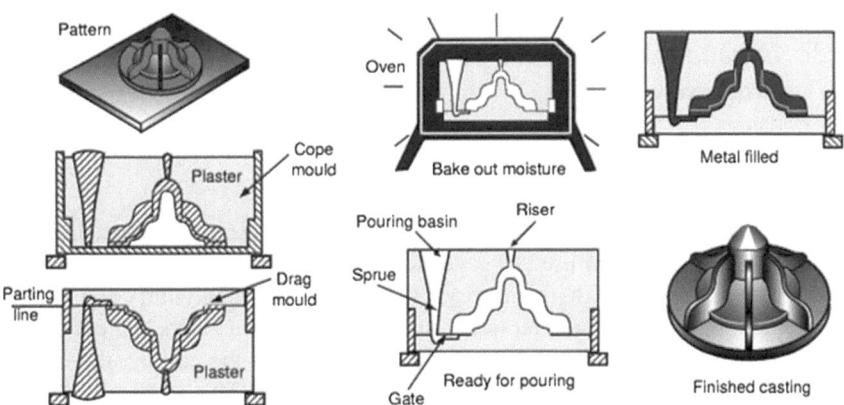

Fig. 8 Plaster mold casting [8]

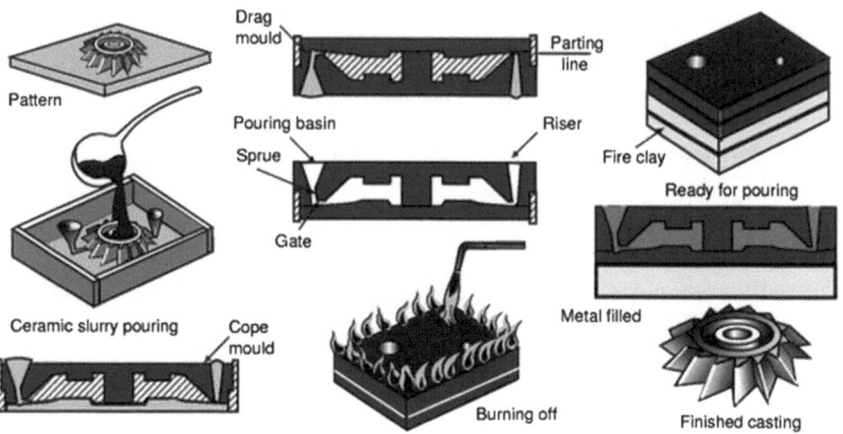

Fig. 9 Ceramic mold casting [8]

3.1.4 Ceramic Mold Casting

In ceramic molding, a fine grain slurry is poured over the pattern and allowed to set chemically. The mold is dried and baked. Molten metal is then poured into the mold and allow to solidify. The casting is removed by breaking the mold. All metals can be cast but problems may arise while casting aluminum, magnesium, and zinc, tin and copper alloys. Some typical applications are all types of dies and molds for different casting and forming processes, cutting tool blanks, component of food handling machines etc. Figure 9 represents the details of a ceramic mold casting process.

3.2 Expendable Mold, Expendable Pattern Casting Processes

3.2.1 Investment Casting

Investment casting process begins with preparing a metal die by either machining or casting. The mold is then used to generate a wax pattern of required shape. Next, the patterns are coated with a refractory material zircon, ceramic slurry and finally a binder, followed by curing in an oven. Upon curing, the wax is melted out and the metal is then cast in the ceramic mold. In order to eject the casting, the mold is destroyed [8, 9]. The process is represented in Fig. 10. This technique is often referred to as "lost-wax casting". Almost all metals can be cast through investment casting. Some applications are jewelry, turbine blades, levers etc.

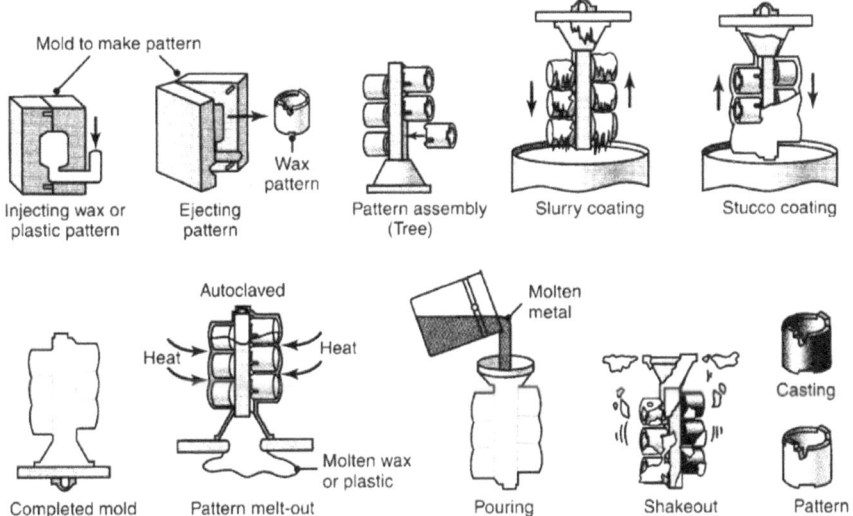

Fig. 10 Investment or lost-wax casting [17]

3.2.2 Evaporative Pattern Casting (Lost-Foam Process)

In evaporative pattern casting, the pattern is prepared from polystyrene. The pattern evaporates as it comes in contact with the molten metal thereby produces a cavity for casting. The process begins with developing a pattern in a metal die. Pattern is then coated with a refractory slurry, dried and placed in a flask as shown in Fig. 11. The flask containing the coated pattern is then filled with either loose, fine sand which surrounds the pattern or with bonded sand for additional strength. After compacting the sand, molten metal is introduced into the mold. The pattern vaporizes and the mold cavity is completely filled with the molten metal. Evaporative casting process is advantageous as the process is simple and economical, and requires minimal finishing and cleaning operations. Also, the process can be automated for long production runs. Typical applications of lost-foam process are similar to investment casting process mentioned above.

3.3 Permanent Mold Casting Processes

3.3.1 Gravity Die Casting

Gravity die casting is often termed as permanent mold casting. In this process, molten metal is poured into a pre-heated die where it solidifies [13]. Cast part is then ejected by opening the die. This casting method is commonly employed for

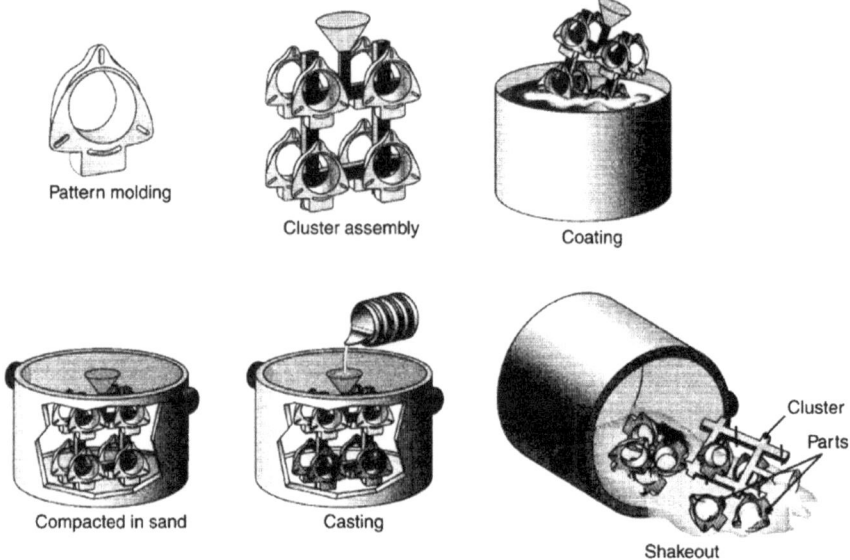

Fig. 11 Evaporative pattern casting [17]

non-ferrous metals such as copper, aluminum, magnesium, but sometimes iron, lead, nickel, tin and zinc alloys. Carbon steel can be cast with graphite dies. Gravity die casting is normally used to produce pistons, kitchen utensils, gear blanks, pipe fittings, and wheel etc. The details of gravity die casting process are presented in Fig. 12.

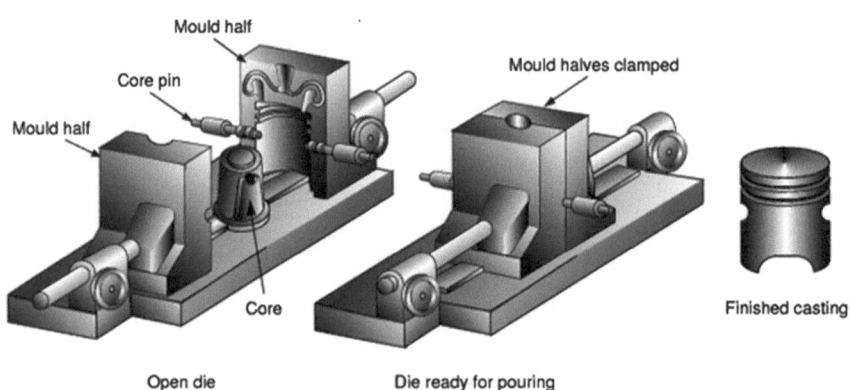

Fig. 12 Gravity die casting process [8]

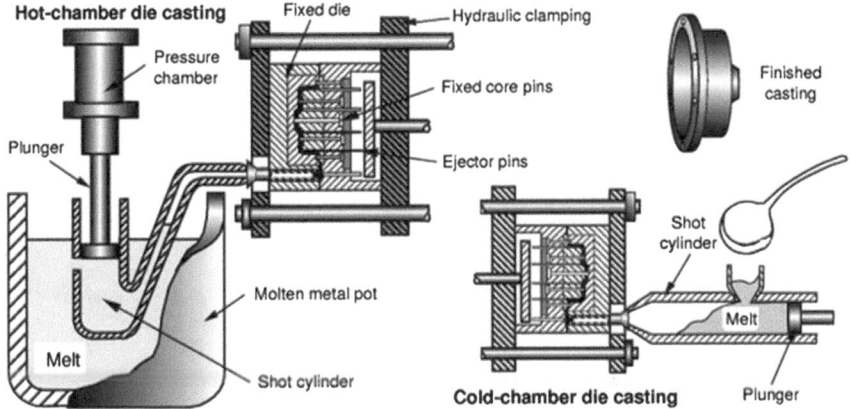

Fig. 13 Pressure die casting [8]

3.3.2 Pressure Die Casting

In high pressure die casting, molten metal is fed into a permanent mold at very high pressures i.e. ~100 bar or more as shown in Fig. 13. The die is remained closed until molten metal completely solidifies, after which the casting is ejected by opening the die. It is important to note that this casting technique is limited to non-ferrous alloys such as zinc, aluminum, magnesium, and lead, tin and copper alloys. Also, feasibility of casting iron based alloys using die casting is under research and development [8]. Some applications of die casting process are domestic appliance components, electrical boxes, pump and impeller parts etc.

3.3.3 Centrifugal Casting

Centrifugal casting method is used to produce cylindrical/tubular parts [4, 6, 8]. During the process, molten metal is poured into a mould which revolves about a vertical or horizontal axis. Vertical axis is often used to cast short work pieces. The rotation speed of the mold depends upon the diameter of the part to be cast and may vary from 300–3000 rpm [8]. All metals suitable for static casting processes can be cast through this technique. An added advantage of this method is the casting of glass, thermoplastics, ceramics and even composite materials. Some of the products of centrifugal casting are pipes, brake drums, engine cylinder liners, nozzles, gun barrels etc. Figure 14 depicts the process details and a typical product prepared by centrifugal casting.

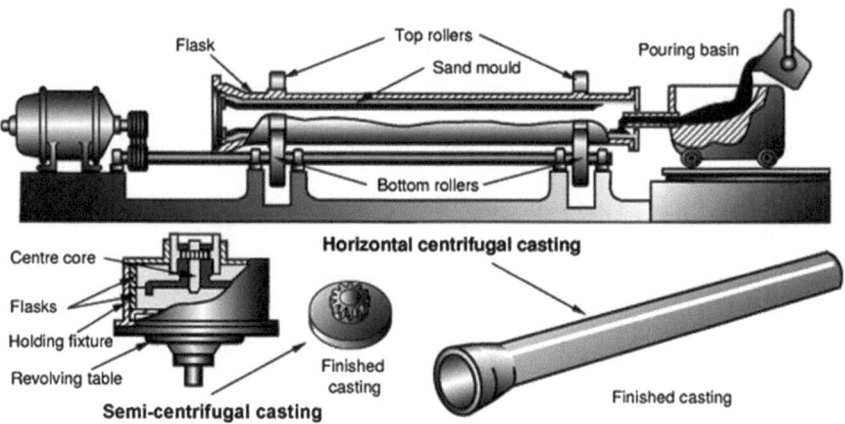

Fig. 14 Centrifugal casting [8]

3.3.4 Squeeze Casting

Squeeze casting is often considered as a combination of casting and forging operation [7]. In this process the molten metal is fed into a preheated mold from the bottom. The top half of the metal applies a high pressure to compress the metal into the final shape of desired product. This technique is also known as "load pressure casting" or "liquid metal forging". Mostly, non-ferrous alloys are cast using this method. Some applications include aerospace components, suspension parts, brake rotors, engine pistons, etc. A schematic diagram of the process is shown in Fig. 15.

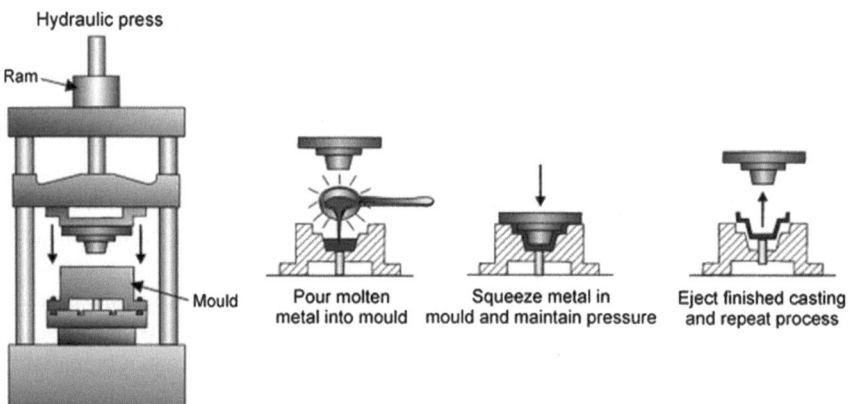

Fig. 15 Squeeze casting [8]

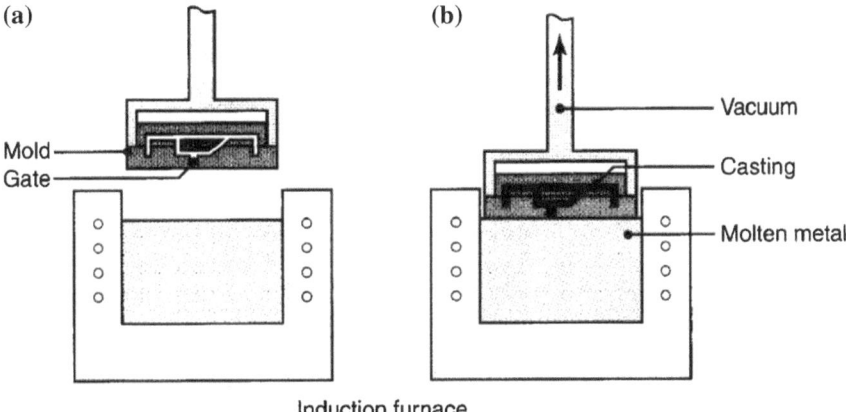

Fig. 16 Vacuum casting. **a** Before and **b** After immersion of mold into molten metal [17]

3.3.5 Vacuum Casting

In vacuum casting process, fine sand and urethane is mixed and molded over metal dies followed by curing with amine vapor. A robotic arm is often used to hold the mold which is then immersed partially in molten metal held in an induction furnace. Vacuum in the mold draws the molten metal through a gate at the bottom of the mold as shown in Fig. 16. Since, the temperature of the melt is 55 °C above the liquidus temperature of the alloy, it starts to solidify within a very short time. The mold is withdrawn from the molten metal once it is completely filled. This process is most suitable for thin castings. Gas-turbine components (up to 0.5 mm thick) are produced by this process.

A summary of general characteristics of above mentioned casting processes is presented in Table 3.

4 Foundry Practices

As discussed above, casting processes differ from each other in terms of molding technique (expendable or permanent), pattern (expendable or permanent), flow of the melt (pressurized or non-pressurized) etc. However, the general characteristics of a typical foundry process can be fairly explained with respect to a central theme, production of sand castings through conventional molding. The essential steps which follows the sequence from casting design to finished product are presented in Fig. 17. Most of the casting processes follow the same sequence with slight modifications in the process. This section is mainly focused on important practices in a foundry setup such as molding techniques, pattern making, casting alloys, furnaces for melt preparation, pouring cleaning, and heat treatment etc.

Table 3 General characteristics of casting processes [16]

	Sand	Shell	Evaporative pattern	Plaster	Investment	Permanent mold	Die	Centrifugal
Cast materials	All	All	All	Nonferrous (Al, Mg, Zn, Cu)	All	All	Nonferrous (Al, Mg, Zn, Cu)	All
Weight (kg)								
Minimum	0.01	0.01	0.01	0.01	0.001	0.1	<0.01	0.01
Maximum	No limit	100+	100+	50+	100+	300	50	5000+
Typical surface finish (R_a in μm)	5–25	1–3	5–25	1–2	0.3–2	2–6	1–2	2–10
Porosity[a]	3–5	4–5	3–5	4–5	5	2–3	3–4	3–4
Shape complexity[a]	1–2	2–3	1–2	1–2	1	2–3	3–4	3–4
Dimensional accuracy[a]	3	2	3	2	1	1	1	3
Section thickness (mm)								
Minimum	3	2	2	1	1	2	0.5	2
Maximum	No limit	–	–	–	75	50	12	100
Typical dimensional tolerance (mm)	1.6–4	±0.003		+0.005–0.010	+0.005	±0.015	±0.001–0.005	0.015
Equipment	3–5	3	2–3	3–5	3–5	2	1	1
Pattern/die	3–5	2–3	2–3	3–5	2–3	2	1	1
Labor	1–3	3	3	1–2	1–2	3	5	5
Typical lead time[b]	Days	Weeks	Weeks	Days	Weeks	Weeks	Weeks to months	Months
Typical production rate[b] (parts/mold-hour)	1–20	5–50	1–20	1–10	1–1000	5–50	2–200	1–1000
Minimum quantity[b]	1	100	500	10	10	1000	10,000	10–10,000

[a]Relative rating from 1 (best) to 5 (worst)

[b]Approximate values without rapid prototyping technologies. Minimum quantity is 1 with the use of rapid prototyping

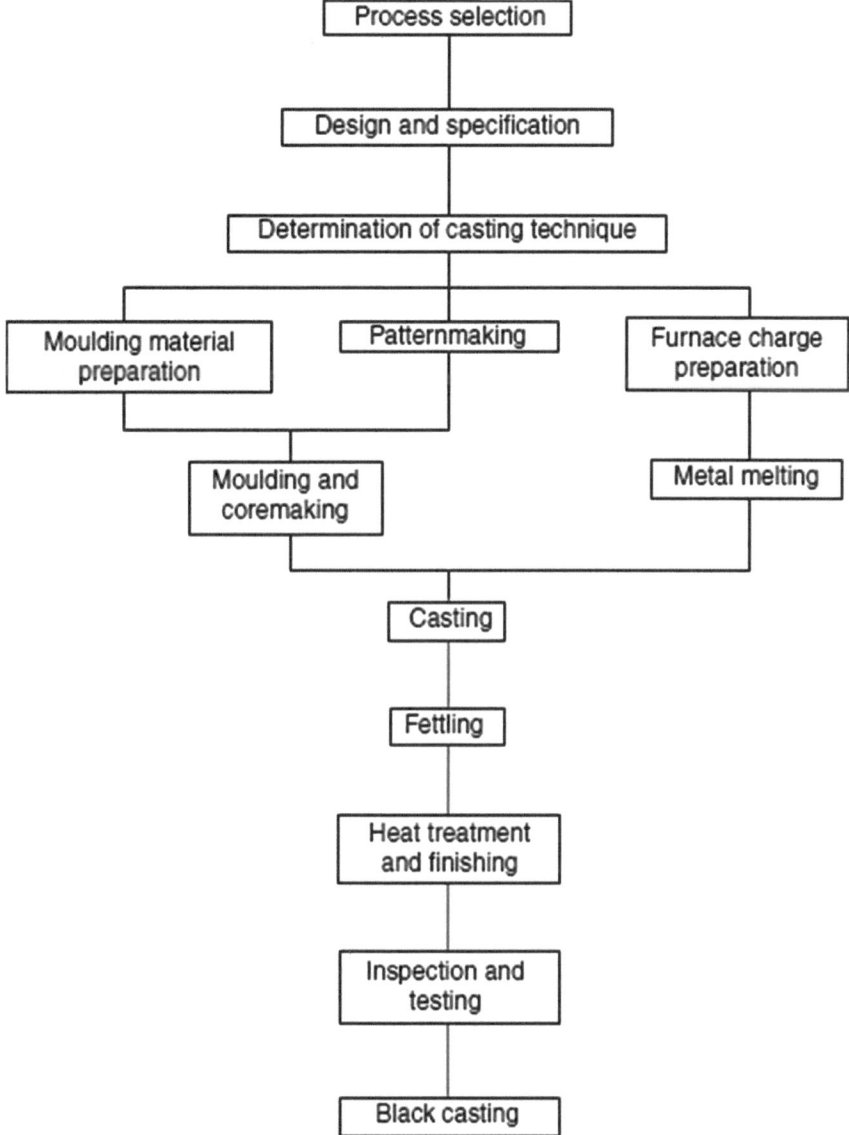

Fig. 17 Casting process flow diagram in a foundry [27]

4.1   Mold Materials and Molding Techniques

Mold materials and molding techniques are exposed to high temperature molten metals during any casting process. It is of utmost importance to carefully select the appropriate mold material and molding technique in order to obtain high quality

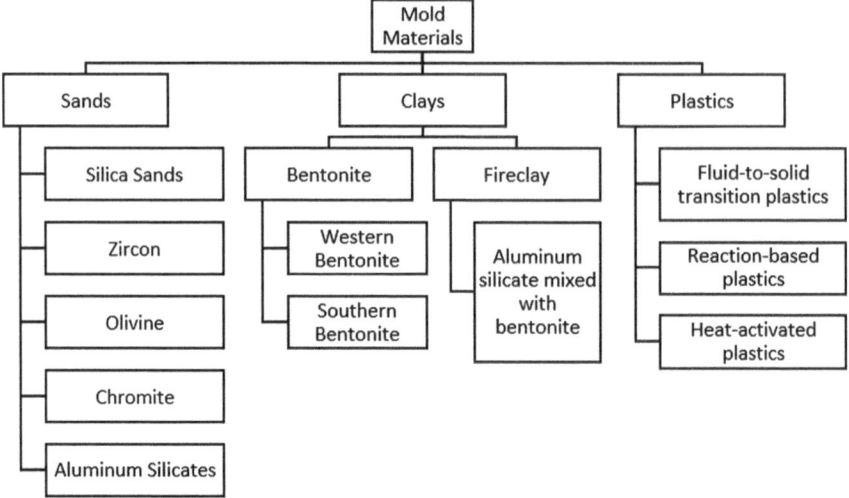

Fig. 18 Mold material used in a foundry

cast products. The use of sand molds in metal casting industry is quite prevalent owing to its low cost in comparison to that of a permanent mold, making it the most suitable choice for low and medium production runs [1]. These molds are generally made up of sands, clay and binders (thermoplastic or thermosetting resins) as shown in Fig. 18. The selection criteria includes but not limited to the type of melt, type of casting, availability of molding materials, mold and core preparation capabilities in foundry, and quality requirements of the consumers.

4.1.1 Sands

Sand is one of the most common refractory material used for mold preparation. A variety of sands are available for this purpose, however, all of them should possess some general properties which are as follows.

(a) High temperature thermal and dimensional stability
(b) Chemically inert with the melt
(c) Suitable size and shape of particles
(d) Not readily wetted by the melt
(e) Economical availability
(f) Free from chemicals which produce gases upon heating
(g) Consistent in terms of cleanliness, composition, and pH
(h) Compatibility with binding agents.

Foundry sands are primarily based on silica (SiO_2) in the form of quartz. Despite the fact that silica sands are cheap and readily available, it poses some challenges during casting operation. An example is the formation of a slag-type compound which produces a rough layer of molten metal and sand on the surface of cast products. However, addition of other materials to silica sand mix can alleviate such problems thereby making it the most widely used molding aggregate in the sand casting industry.

Porosity is another important consideration as it directly influences the permeability of the mold, which refers to the ability of the mold to release the gases generated during pouring and mold filling. A sand with uniform grain size offers more porosity compared to the one with distributed grain size. It is easier for small grains to take spaces in between the large grains and thus making the mold less permeable. Nevertheless, porosity should not be too high to allow penetration of the melt into sand grains resulting in a burn-in defect. Hence, a balanced sand distribution is necessary to optimize the molds for their porosity and permeability.

Zirconium silicate or Zircon ($ZrSiO_4$) is another sand with very good foundry characteristics such as low thermal expansion, high thermal conductivity and bulk density making its chilling rate four time to that of quartz, chemical inertness with the melt. This sand is commonly used in casting steel and investment casting process for high-temperature alloy products [14]. Olivine sands are composed of forsterite (Mg_2SiO_4) and fayalite (Fe_2SiO_4) and provides different physical properties based on their compositions. Silica and olivine sands have similar specific heat but olivine sands offer low thermal expansion and thus, are primarily used to control the mold dimensions in steel castings. Also, olivine sands are somewhat less durable than silica sands.

Another commonly used sand in foundry is Chromite ($FeCr_2O_4$) which is thermally stable, chemically unreactive, and with good chilling properties. Besides its higher thermal expansion than zircon sands, it causes pinholes and gas defects in cast products due to hydrous impurities within this sand. The use of aluminum silicate (Al_2SiO_5) in the form of kyanite, sillimanite, and andalusite is also observed in a foundry. Some of the characteristics of this sand are high refractoriness, low thermal expansion, and resistance to thermal shock. Aluminum silicate in combination with zircon are commonly used in precision investment foundries.

4.1.2 Clays

The interaction of clay and water results in bonding a green sand mold. Each of these clay offer different properties based on its composition and structure. One example of these clays is bentonite which is most widely used in foundries. It is a layered structure and it produces clay particles in the form of flat plates. The absorption of water takes place on these plates due to which bentonite expands and contracts in the presence and absence of water respectively. Bentonite is further classified in two forms which are Western (sodium) bentonite and Southern (calcium) bentonite.

Western bentonite possess high green and dry strengths making it suitable for ferrous alloys. Also, it offers high degree of plasticity, toughness, and deformation together with excellent frictional properties when mulled with water. Molding sand with western bentonite sets well around pattern resulting in the higher strength of the mold. In addition to that western bentonite can swell up to 13 times its original volume [1], making it an appropriate agent between sand grains after compaction in the mold. Defects due to sand expansion could be easily reduced using this molding sand-clay combination. An added advantage of western bentonite is its durability which enables it to be used multiple times with minimum bonding additions. It is important to control the clay/water ratio while using western bentonite to ensure good shakeout characteristics of the mold. Southern bentonite, on the other hand, is lower-swelling clay which provides higher green compressive strength and lower dry compressive strength (~ 30–40 %) as compared to western bentonite. Shakeout characteristics are good owing to lower hot compressive strength. The easiness in flow of southern bentonite bonded sands make it a choice for preparing the molds which contains pattern with intricate details such as pockets and crevices.

Bentonite mixed with a hydrous aluminum silicate kaolinite results in fireclay which shows high refractory and low plasticity characteristics. This addition to bentonite allows to vary water content over a great range and also improves hot strength of the mold, thereby making it suitable for large castings. The use of fireclay is limited owing to the durability issues. Also, close control of sand mixes and materials can completely eliminate the use of fireclay in mold preparation.

4.1.3 Plastics

Plastics materials (or commonly known as resins) are extensively used as binders in metal casting industry. Resins are normally used in binding sands and core manufacturing for all sizes and production volumes. Plastics are broadly categorized in three categories: Fluid-to-solid transition plastics, Reaction-based plastics, and Heat-activated plastics. Fluid-to-solid transition plastics are composed of liquid polymeric binders which cross link and set up in the presence of a catalyst, thus, transforms from a liquid to a solid. Reaction-based plastics forms a solid polymeric structure in the presence of a catalyst. Heat-activated plastics are thermoplastics or thermosetting resins which are added to the sand as dry powders followed by heating of the mixture. As a result, a thermosetting reaction takes place when the powder melts and flows over the sand. Alternatively, a heat activated reaction may take place between two liquids to form a solid. Selection of any binder system is critical because these systems are often sensitive to temperature and humidity. Also, environmental issues must be considered as some of the binders emits noxious odors and fumes.

4.2 Patterns and Cores

Pattern making is one the crucial steps towards a "good" casting. Pattern are an integral component of a casting process as molding material (usually prepared sand) is formed around it to shape the casting cavity. Patterns are prepared using plastics, wood, or even metal depending on the complexity of the part to be cast and the number of cast products required in a batch. Some of the important considerations in pattern design are as follows.

1. Pattern should have an allowance for solid state shrinkage which is an inherent feature of any casting during solidification from melting temperature to room temperature.
2. A draft angle should be provided such that the pattern can be removed smoothly from the mold without destroying it.
3. Inclusion of enough extra stock to take into account the variations that might arise in casting dimensions due to mold preparation, pattern wear etc.

4.2.1 Pattern Types

Patterns are design in a number of ways to meet the process requirements and economic considerations [17]. Selection of a particular pattern type is based on number of cast products required, molding and casting technique employed, pattern size, and casting tolerances needed. Some patterns are reusable over a large number of production cycles of sand castings such as solid pattern, split pattern, matchplate pattern, and cope-and-drag patterns as shown in Fig. 19. Besides these commonly used patterns for sand castings, there are special patterns made up of metals and/or low melting temperature substances for various other casting processes. The details of most commonly used patterns in a typical metalcasting industry are as follows.

Solid Pattern: Solid pattern is the simplest and cheapest of all pattern types. It is made up of a single piece and based on its size and shape, it could be molded in one or two boxes. Molding of a solid pattern requires a lot of manual operations due to

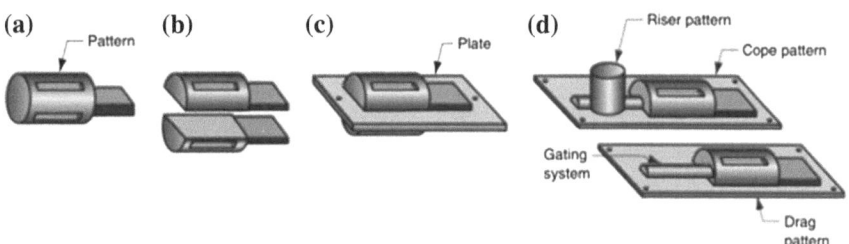

Fig. 19 a Solid pattern. **b** Split pattern. **c** Matchplate pattern and **d** Cope-and-drag pattern [6]

which its use is limited to casting few components. In general, mold is prepared by placing the solid pattern in either cope or drag portion of the mold.

Split Pattern: The production of intricate casting necessitates the pattern to be split in two parts. For such casting, the pattern is divided into two halves as shown in Fig. 19b along a plane coincident with the parting line of the mold.

Matchplate Pattern: A matchplate pattern is formed by attaching the two pieces of a split pattern to opposite faces of a metal or wood plate. The two holes in the plate as shown in Fig. 19c allows perfect alignment of the top and bottom portion of the resulting mold. Matchplate patterns are often used to produce small and accurate casting on a large scale with minimum production time.

Cope-and-drag Pattern: These pattern are similar to the matchplate pattern except that the split pattern halves are attached to different plates in this case. Cope and drag portions of the mold are prepared independently and matched using dowel pins. Gating and riser system also have to be incorporated within the pattern design while using cope-and-drag pattern as shown in Fig. 19d.

Gated Pattern: This type of patterns is used in mass production of castings in minimum production time. The pattern is design in such a way that it produces a multi-cavity mold for casting similar parts as shown in Fig. 20a.

Skeleton Pattern: Skeleton patterns are used to produce heavy and large castings in a small quantity. This pattern type is economically appealing as making a solid pattern for large cast parts is expensive. A skeleton pattern consists of wooden ribs mounted on a metal base which provides a framework for casting. The pattern is filled with rammed sand followed by removal of excess sand through strickleboard as shown in Fig. 20b [12]. This type of pattern is normally used for casting turbines, L-bends and water pipes etc.

Sweep Pattern: A sweep pattern is used to cast cylindrical objects which are symmetrical about an axis. Main components of this pattern are a base (placed in sand mass), a vertical spindle, and a wooden template as shown in Fig. 20c. The outer edge of the template contains a contour which corresponds to the shape of required casting. Upon rotating the vertical spindle, a cavity is produced inside the rammed sand.

Segmental Pattern: Segmental patterns are usually a segment of a final cast product. This pattern type is popular for making circular components such as rims, wheels, gears, pulleys etc. Figure 20d represents a typical segmental pattern.

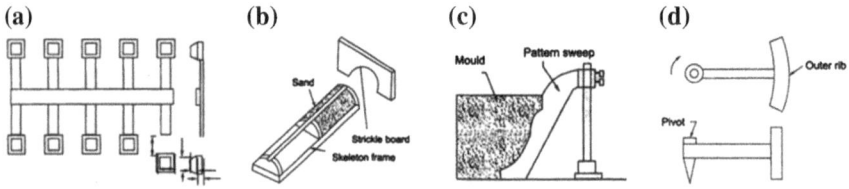

Fig. 20 **a** Gated pattern. **b** Skeleton pattern. **c** Sweep pattern. **d** Segmental pattern [12]

Expendable Pattern: The expendable patterns are prepared from low-melting temperature materials such as wax or expanded polystyrene (EPS). These patterns are relatively inexpensive. They are not reusable patterns as they are consumed during the casting process. This type of pattern is used for investment casting and lost-foam process.

4.2.2 Cores

The external shape of the casting can be defined using a pattern as discussed in the section above. If the casting is hollow or it has an inner surface profile, then it requires a core. A core is a full scale model of the interior surface of the cast part [6, 17]. It is an integral part of molds for casting hollow products. During pouring, the molten metal flows between the mold cavity and the core, resulting in the desired hollow casting. The cores are usually prepared with chemically bonded sands and includes shrinkage and machining allowances similar to a pattern. A core can be easily placed inside the mold assembly without any support, however if needed, chaplets (made up of a metal with higher melting temperature as compared to cast metal) are used to hold the core at appropriate position within the mold. Figure 21 shows a mold assembly with a core that is supported by chaplets. Any portion of chaplet that protrudes on the final casting is removed during finishing operations.

4.2.3 Pattern-Less Casting Technology

In more recent years a new technology of additive manufacturing has been developed in order to reduce the casting time and cost. Such techniques are often termed as "Rapid Casting" processes [24], where no pattern or die is needed. Instead, the mold is prepared by milling in one time. During molding, sand molds and cores can be integrated to reduce the number of sand cores, resulting in a

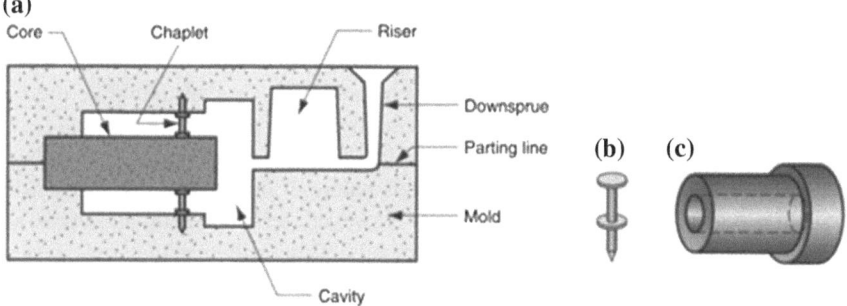

Fig. 21 a Core in a mold supported by chaplets. **b** A chaplet design. **c** Hollow cast product [17]

simplified design and reduced allowance of machining [15]. Casting dimension are easy to control in this method. The mold cavity and cores can be synchronously shaped which improves the accuracy of sand mold and casting. Since, no pattern draft is needed, the final castings are found to be less in weight compared to conventional castings. Castings of intricate designs became possible with this method, especially the accuracy of curved surfaces can be maintained and controlled to a great extent.

One of the most promising technologies in metal casting is three dimensional printing, often known as 3D-printing. This technique has gained popularity in recent past and has resulted in significant improvements in processing speed and minimizing cost [3]. The 3D printed prototypes can be used to directly produce the molds for casting or used as a pattern to produce mold indirectly. Another improvement in the pattern less casting technology is ZCast process where the complex cavities and cores are 3D-printed using a ceramic material. ZCast process is limited to cast only light alloys. The following sub-section provides insights into these two rapid casting technologies.

Three Dimensional (3D) Printing

Three-dimensional (3D) printing is the process of joining material, layer-by-layer, to make objects from 3D model data (usually created by a computer-aided design software or a scan of an existing object) [3]. The parts are built upon a platform situated in a bin full of powder material as shown in Fig. 22a. A layer of powdered material is distributed at a time which is then hardened and joined together by depositing the drops of binder in a manner similar to inkjet printing [24]. A piston is used to lower the part so that the next layer of powder can be applied. For each layer, the powder hopper and roller system distribute a thin layer of powder over the top of the work tray. The cross-section of the part is selectively hardened by applying binder through continuous jet printing nozzles during a raster scan of the work area. The process is repeated and the subsequent layers are applied using the loose powder which was unbound previously. Upon completion, the part is removed from unbound powder and the excess unbound powder is blown away. For durability and a good surface finish, the part may be coated with wax, CA glue or any other sealant. A complete cycle of 3D printing process is shown in Fig. 22b.

A range of materials can be used in 3D printing. Any powdered material with a binder of sufficiently low viscosity capable of producing droplets could be used in 3D printing of objects [20]. Not only ceramics but plastics, metal and even metal-ceramic composites can also be produced by this technique. A drawback associated with 3D printing is that the porous nature of products due to the density limitation on the dry powder distribution. However, ceramic shells and cores are prepared and used frequently for casting metals. Three different materials system that have been developed for 3D printing systems [24], are as follows.

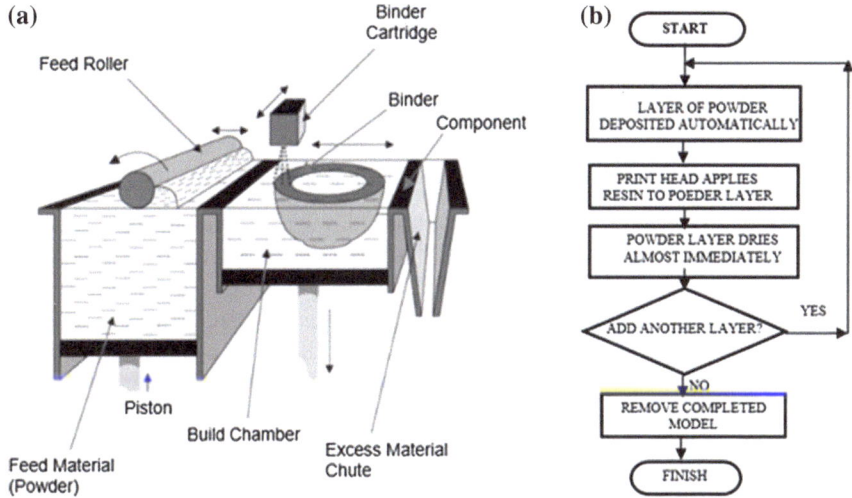

(a)

Binder
Cartridge

Feed Roller

Binder

Component

Piston

Build Chamber

Feed Material
(Powder)

Excess Material
Chute

(b)

START

LAYER OF POWDER
DEPOSITED AUTOMATICALLY

PRINT HEAD APPLIES
RESIN TO POEDER LAYER

POWDER LAYER DRIES
ALMOST IMMEDIATELY

ADD ANOTHER LAYER?

YES

NO

REMOVE COMPLETED
MODEL

FINISH

Fig. 22 **a** 3D printing process. **b** Process flow diagram of 3D printing [21, 24]

Plaster Based Material: It is a powder/binder system which consists of an oxidant and reductant (a redox couple). Upon interaction of binder and powder, an acid is produced which accelerates cross-linking. The strength of 3D printed product is increased y this mechanism. The oxidant may be in powder and the reductant in the binder or vice versa. Some characteristics of plaster based materials are high strength, ability to produce delicate and thin walled parts, accurate representation of design details, and color printing [24].

Composite Based Materials: These materials are suitable for thin-walled enclosures and assembly applications.

Starch Based Materials: This system uses starch based polymer powders such as cornstarch, dextran and gelatin for 3D printing process. Some characteristics of these systems are high speed printing, ability to print large bulky parts, and printing of patterns for investment casting [24].

ZCastTM Direct Metal Casting

ZCast direct metal casting is one the rapid casting techniques developed primarily for the casting of non-ferrous alloys. It creates the shell molds directly from CAD model using 3D printing. In contrast to conventional metal casting which is constrained in dimensional accuracy by pattern extractability, the layer by layer construction allows creating complex objects, without restricting the undercuts provided that the unconsolidated powder can be removed from the cavity [5]. This method simply eliminates the pattern creation phase in conventional casting and thus reduces production time from weeks to days [25]. ZCast provides three basic methods to manufacture molds for rapid casting. It was reported that accuracy and surface finish are consistent with that of sand casting [2]. Some of the major features of ZCast process are it is recommended for non-ferrous metals with pouring temperature below 1100 °C. The shell mold wall thickness ranges from 12.5 to 25.4 mm. The ZCast molds should be baked in oven for 4–8 h before used in metal casting process [24]. The three ZCast methods are explained below.

Direct Metal Casting (ZCast): This method allows to create molds (copes and drags) and cores directly from a CAD file. No pattern is required resulting in reduced production time. The process includes designing of basic parting line and coring. A 3D mold shell is printed using ZCast plaster-ceramic composite material. If needed, the shell is created with ribbing and backfilled with conventional foundry sand to improve strength at minimum material cost.

Loose Pattern Method (LP): The loose pattern method is an established method in casting industry. With loose pattern method, the patterns are prepared using 3D printing. A split line has to be created around the pattern by using resin mixed foundry sand or similar materials. The pattern stays loosely seated so that the process of creating the split line can be repeated for the other mold half, using the same pattern. The two joint boards are framed with wood together with 3D printed pattern constitutes the final foundry tool. Mold halves are then created separately by interchanging the pattern between the two core boxes.

Production Intent Casting (PIC): This method is a combination of creating patterns by ZP102 plaster material and ZCast material for creating cores. At first, the foundry tooling is created in a similar way as it is designed for a production foundry process including core prints, offset partings, and clearances. Afterwards, the pattern is printed, coated with epoxy, and backfilled with a rigid plastic filler for improved strength. Cores, if required, are produce by ZCast material.

4.3 Casting Alloys

The development in casting techniques enable to cast many alloy compositions. These alloys can be broadly categorized in two categories: Nonferrous Alloys and Ferrous Alloys as shown in Fig. 23.

Nonferrous Alloys

Aluminum-based Alloys: The hardening mechanisms and heat treatments for aluminum alloys offer a range of mechanical properties which makes it suitable for casting purposes. These alloys are good at corrosion resistance, non-toxic, light in weight, and have good machinability [17]. Wear and abrasion resistance is also good if silicon is not used in alloy compositions. Due to high strength to weight ratio, aluminum castings (also known as light-metal castings) are observed in the

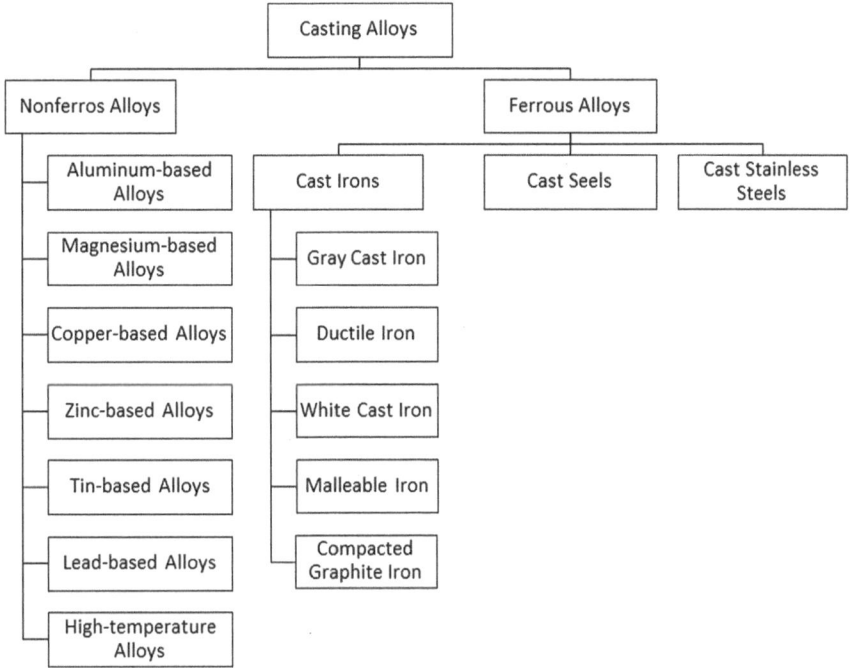

Fig. 23 Classification of casting alloys

form of cylinder heads, transmissions cases, engine blocks, suspension parts, brakes and wheels etc.

Magnesium-based Alloys: Magnesium-based casting alloys are also popular for their lowest density. Although corrosion resistance is good, their strength is determined by the heat treatment done on the alloy. Some examples of magnesium castings include air-cooled engine blocks, housings, and automotive wheels etc.

Copper-based Alloys: Copper-based alloys are advantageous in terms of high thermal and electrical conductivity, non-toxicity, and resistance to corrosion and wear for bearing materials. Although a range of copper-based alloys such as brass, aluminum bronze, phosphorus bronze, tin bronze etc. are available, their use is limited in castings due to higher costs.

Zinc-based Alloys: Zinc-based alloys are common for casting intricate shapes and thin-walled components. These alloys belong to low melting temperature alloy family and shows good corrosion resistance, high fluidity and strength when used in structural applications.

Tin-based Alloys: These alloys performs well in corrosive environments. With lower strength they are not common in large structural applications. Most commonly they are used for casting bearing surfaces.

Lead-based Alloys: The applications of lead-based alloys are similar to tin-based alloys. These are also low-melting temperature alloys. The use of lead-based alloys is limited in application due to toxicity and carcinogenic effects.

High-temperature Alloys: Some alloys require high temperature for casting. A temperature of 1650 °C is required to cast titanium and even higher for other refractory alloys such as molybdenum, niobium, tungsten and tantalum. Typical applications of these high temperature alloys are in casting nozzles and various jet- and rocket-engine components.

4.3.1 Ferrous Alloys

Cast Irons: The use of cast irons in metal casting is widespread due to a number of properties such as easy to cast intricate shapes, hardness, and good machinability and wear resistance. Cast iron refers to a family of alloys, the details of which are presented as follows.

(a) Gray Cast Iron: Gray cast iron is available in different forms such as pearlite, ferrite, and martensite. Each of these forms has different properties which is attributed to their microstructures. Some of the applications include engine blocks, motor housings, pipes, and wear surfaces for machines.

(b) Ductile (nodular) Iron: Ductile irons are used for casting various machine parts such as gears, automotive crankshafts, pipes, rolls for rolling machines, housings etc. They are specified by a special set of two digits numbers. For example 75–50–06 represents that the material has a minimum tensile strength of 75 ksi, a minimum yield strength of 50 ksi, and 6 % elongation in 2 in.

(c) White Cast Iron: These irons offer excellent wear resistance together with high hardness. Most common applications include railroad car brake shoes, rolls for rolling mills, and liners in machinery for processing abrasive materials.

(d) Malleable Iron: This iron is utilized in manufacturing or railroad equipment, fittings, and parts for electrical applications.

(e) Compacted Graphite Iron (CGI): The properties of CGI lies somewhere between gray and ductile irons. CGI has good damping and thermal properties together with high strength and stiffness. It is usually employed for casting smaller parts with higher strength. The machinability is also good as compared to ductile iron which makes it a good choice for casting automotive engine blocks and cylinder heads.

Cast Steels: The use of cast steel requires very high temperatures (up to 1650 °C) due to which mold material has to be selected after careful consideration. However, casting steel results in more uniform properties in a part as compared to any other metal working process [17]. Cast steels can also be welded to join separately cast products, however, it alters the microstructure in the heat affected zone. Welding must be followed by an appropriate heat treatment so that the properties of casting could be restored. Typical applications of cast steel include but not limited to mining, chemical plants, heavy constructions, oil fields, and railroad equipment. *Cast Stainless Steels*: Stainless steels show long freezing ranges and very high melting temperatures [17]. These steels are treated in a similar manner as of cast steels. Based on composition and processing parameters, these steels produce different microstructures. Cast steel alloys are good for heat and corrosion resistance, especially nickel-based casting alloys are the most suitable in extremely corrosive environments and in very high temperature applications.

4.4 Furnaces and Melting Practice

The melting of cast metals is an essential step in any casting process. This requires heat input which is supplied through a furnace. These furnaces are charged with melting stock which consists of metal, alloying elements, and other materials such as flux and slag-forming products. The function of fluxes is to remove impurities and dissolved gases from the melt to be poured in mold cavity. Some of the commonly used furnaces in a foundry are discussed below.

Electric Arc Furnaces: An electric arc is used to melt the charge in this type of furnaces. Electrical arc furnaces are design in different configurations (two or three electrodes) as shown in Fig. 24. They consume a lot of power and are used for

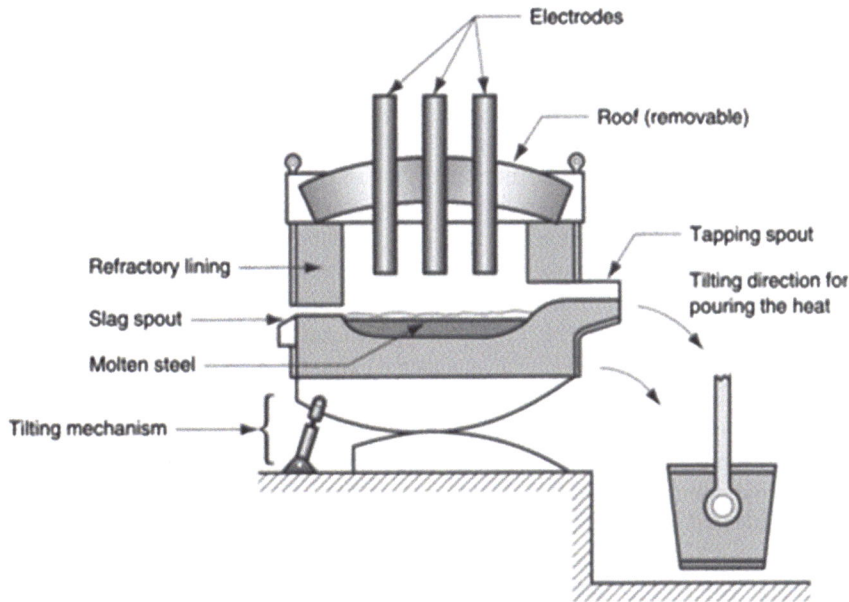

Fig. 24 Electric arc furnace for casting steel [6]

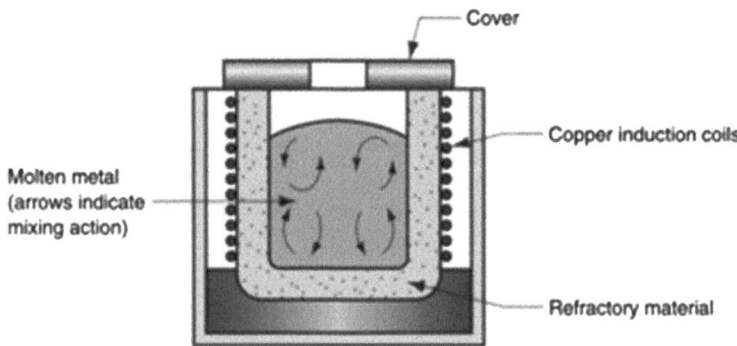

Fig. 25 Induction furnace [6]

higher melting rates up to 23,000–45,000 kg/h [17]. These furnaces are less polluted and can hold the melt at a particular temperature for a period of time for alloying purposes. Electrical arc furnace are mainly used for casting steels.

Induction Furnaces: Induction furnaces are used in small foundries and are most suitable for smaller quantities and controlled melt compositions. The principle is relatively simple where alternating current passes through a coil to develop a magnetic field, and the resulting induced current rapidly heats and melts the metal. A simple inductions furnace is shown in Fig. 25. There are two main types of induction furnaces: Coreless Induction Furnace and Core or Channel Furnace [17]. In coreless furnace, a high frequency current is passed through the copper coils which encloses the crucible. The mixing characteristics are excellent due to a very strong electromagnetic stirring action during induction heating. Core or Channel furnace, on the other hand, contains a portion of the unit which is surrounded with coils through which a low frequency (~ 60 Hz) current flows. Some advantages of induction furnaces are superheating which improves fluidity, holding, and duplexing which allows transferring melt from one furnace to the other. Melts from steels, cast irons, and aluminum alloys can be prepared using induction furnaces.

Crucible Furnaces: In crucible furnaces, molten metal and burning fuel mixture do not have a direct contact, therefore, they are often terms as indirect fuel-fired furnaces. There are three main types of crucible furnaces: lift-out type, stationary and tilting as shown in Fig. 26. In lift-out type, a crucible is placed in a furnace and heated sufficiently to melt the molten material. These furnaces are powered by oil, gas, or powdered coal. Once the melt is prepared, the crucible is lifted out from the furnace and used as a pouring ladle. In other two types, the container and heating furnace are integrated due to which they are also termed as pot furnaces. In stationary configuration, the molten metal is ladled out from the container, however, in tilted configuration the whole assembly can rotate to fill the mold. Crucible furnaces are employed in melting non-ferrous metals such as aluminum and zinc alloys, brass, and bronze etc. and are limited to be used for several hundred pounds of metal.

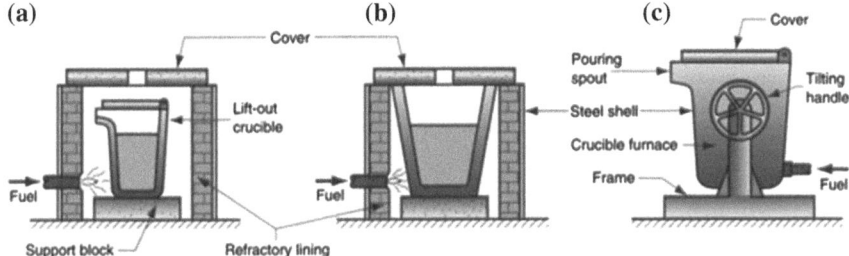

Fig. 26 Crucible furnaces: **a** Lift-out type. **b** Stationary. **c** Tilting [6]

Cupola Furnace: These furnaces are vertical units made up of steel with an inner refractory lining. The main advantages of using a cupola furnace are their continuous operation, higher rates of melting, and large molten metal volumes as compared to any other type of furnace. Figure 27 shows schematic diagram of a cupola furnace. The charge (metal, coke and flux) is loaded through a charging door located in the middle of the furnace assembly and it is charged in alternating layers within the vessel. Iron is usually a combination of pig iron and scrap from previous castings, coke is used as fuel for heating the furnace, and the flux is limestone which reacts with impurities and coke ash to form slag (to avoid heat loss and to keep molten metal non-reactive with the environment). The combustion of coke takes place by forcing the air through openings near the bottom of the furnace. As the heating continues and melt prepares, the furnace is tapped periodically to extract the molten metal for pouring.

Direct Fuel-Fired Furnaces: In a direct fuel-fired furnace, the metal is heated in a small open-hearth by fuel (natural gas and combustion products) burners located at the sides of the furnace. The flame is reflected by the roof of the furnace and thus

Fig. 27 Cupola furnace [17]

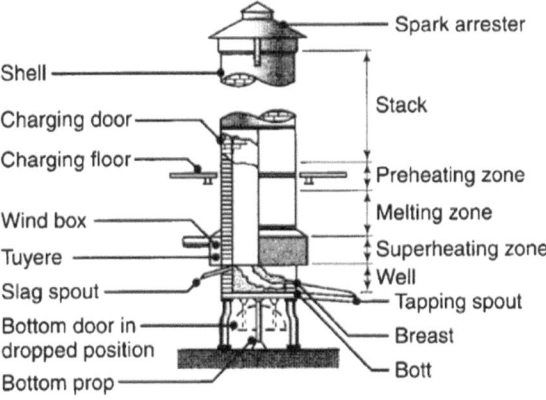

enhance the heating effect for melt preparation. Melt is than extracted through a tap hole at the bottom of the furnace. Typical applications of this furnace type is in casting nonferrous alloys such as aluminum- and copper-based alloys.

4.5 Pouring, Cleaning and Heat Treatment

The movement of molten metal from furnace to the mold is a critical step in a casting foundry due to very higher temperatures of the melt. This is usually done by ladles, a common name for containers that contain melts after taking it out from the furnace. Different type of ladles are available in a foundry most common of which are crane ladle and a two-man ladle as shown in Fig. 28. Crane ladles are used for carrying large amount of molten metals and their movement is controlled by a crane. Conversely, a two-man ladle contains small amount of melt and it is operated manually. A problem faced during pouring is the introduction of oxides into the mold cavity which results in casting defects. In order to avoid such occurrences either fluxes are used to avoid oxidation or filters are used to catch the oxides during pouring. An alternative solution to this problem is to allow flow of molten metal in the mold from bottom of the ladle since the top surface is usually accumulated with the oxides.

Once the molten metal is poured and solidified in the mold cavity, a number of additional operations have to be done to obtain a finished cast product. These additional operations often referred to as *Cleaning* in a foundry which include trimming, removal of cores (if added in mold), cleaning of cast surface, inspection for quality check, and repair if needed. The removal of all other components attached to cast parts such as sprue, runners, risers, parting line flash, fins, chaplets etc. is called *trimming*. Trimming may require hammering, shearing, hack-sawing, band-sawing, abrasive wheel cutting or any other torch cutting methods. If final

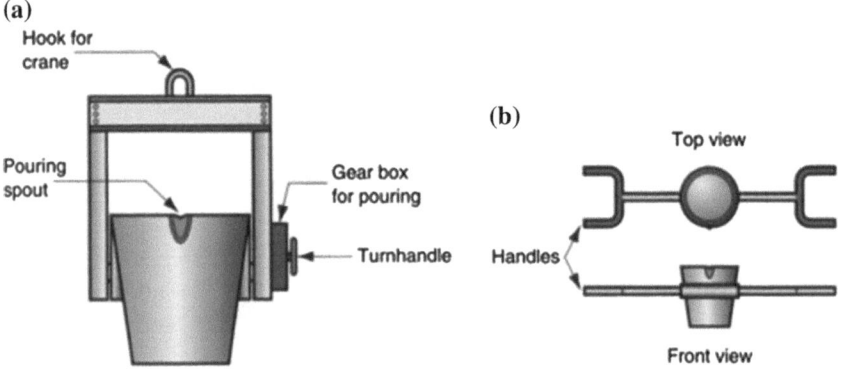

Fig. 28 a Crane ladle and b Two-man ladle for pouring [6]

casting contains cores, they are removed either by manual or mechanical shaking or in some instances the bonding agent in core is dissolved chemically.

Surface cleaning of the final cast product is important, the extent of which depends on the process used for casting. Sand castings need more cleaning whereas castings from permanent mold processes requires minimal or even no cleaning at all. Some cleaning methods for sand castings include wire brushing, buffing, air-blasting with metal shot or coarse sand grits, tumbling, and chemical pickling etc. The quality of casting is then evaluated during inspection through various techniques during which all casting defects are identified and repaired if required. Last of all, castings are often subjected to *heat treatment* either to improve properties for subsequent machining operations or to obtain required properties for application of cast parts.

5   Casting Defects

The process of casting involves a host of parameters which increases the probability of having defects in the final cast product. These defects may arise due to cast material, product geometry, and process techniques. While some of the casting defects are visual which affect the appearance of product, others can have adverse effects on the performance of castings in service. It is, therefore, important to analyze these defects together with their root causes and possible remedies. Casting defects can be classified into four categories as (a) filling related defects (b) shape related defects (c) thermal defects and (d) defects by appearance. Figure 29 shows various types of defects under each of these categories. Possible causes and remedies for different defects are listed in Table 4.

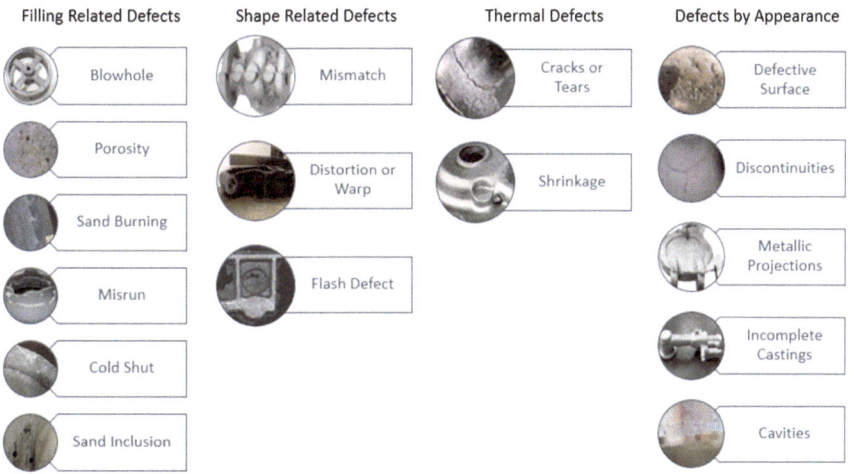

Fig. 29 Types of casting defects

Table 4 Casting defects, causes and remedies [22]

Casting defect		Definition	Causes	Remedies
Filling related defects	Blowhole	Cavity defect formed by gases entrapped during solidification	(1) Inadequate venting (2) Low gas permeability of sand	(1) Improve venting (2) Use coarse sands for improved gas permeability
	Sand burning	Sticking sand to the casting at higher temperatures	(1) Uneven mold compaction (2) Very high melt temperature (3) Uneven distribution of inflowing metal	(1) Ensure uniform compaction (2) Even out incoming metal flow (3) Reduce pouring rate
	Sand inclusion	Slag inside of metal castings	(1) Uneven mold compaction (2) Pouring rate too high (3) Ladle too far above pouring basin (4) Pouring time too long	(1) Ensure uniform compaction (2) Avoid high pouring rates (3) Shorten pouring time (4) Improve distribution of gates
	Cold shut	When two metal streams do not fuse together properly due to poor gating system	(1) Lack of fluidity in melt (2) Faulty design (3) Faulty gating	(1) Adjust proper pouring temperature (2) Modify design (3) Modify gating system
	Misrun	Incomplete casting defect when metal is unable to fill complete mold cavity	(1) Lack of fluidity in melt (2) Faulty design (3) Faulty gating	(1) Adjust proper pouring temperature (2) Modify design (3) Modify gating system
	Porosity	Defect arises due to air entrapment within the mold	(1) Low metal pouring temperature (2) Lack of fluidity in melt (3) Pouring too slow (4) Interrupted pouring	(1) Increase metal pouring temperature (2) Improved melt fluidity (3) Fast pouring (4) Adequate venting of molds and cores
Shape related defects	Mismatch defect	Defect due to shifting molding flashes. It causes the dislocation at parting line	(1) Improper positioning of cope and drag (2) Loose box pins, inaccurate pattern dowel pins	(1) Check pattern mounting on match plate rectify, correct dowels (2) Proper molding box and closing pins
	Distortion or warp	Distortion due to warpage	Residual stresses in casting	Heat treatment
	Flash defect	Unwanted, excess material which forms at parting surfaces	(1) Bending, crowning or stretching of dies (2) Insufficient machine clamp-up (3) Cavities offset from center of plate	Weight down the mold

(continued)

Table 4 (continued)

Casting defect		Definition	Causes	Remedies
Thermal defect	Cracks or tears	Lines on the surface of castings which separates it without breaking	(1) Shrinkage of casting (2) Uneven or excessive ejection forces (3) Insufficient draft (4) Excessive porosity	(1) Reduce pouring temperature (2) Avoid superheating of melt (3) Use chills and proper feeders (4) Avoid early knockout (5) Reduce sharp corners
	Shrinkage	When feed metal is not available to compensate for shrinkage during solidification	Difference in alloy density in molten and solid state	Ensure liquid metal under pressure continues to flow into the voids as they form
Defects by appearance	Metallic projection	Joint flash or fins	(1) Clearance between two elements of the mold (2) Poorly fit mold joint	Care in core setting and mold assembly
	Cavities	Blowholes, pinholes, smooth-walled cavities	Gas entrapment during solidification	(1) Appropriate venting (2) Increased gas permeability of mold and cores
	Discontinuities	Hot cracking	(1) Rough handling (2) Excessive temperature at shakeout	(1) Care in shakeout (2) Proper handling (3) Sufficient cooling in mold
	Incorrect dimension or shape	Distorted castings due to improper ramming of mold	Insufficient rigidity of pattern or pattern plate to withstand the ramming pressure applied to the sand	Assure adequate rigidity of patterns and pattern plate
	Defective surface	Flow marks	Oxide films which lodge at the surface	(1) Increased mold temperature (2) Lower pouring temperature (3) Modify gating system

5.1 Filling Related Defects

Blowhole: This defect arises due to entrapment of gases during solidification process. A round or oval cavity is formed either at the surface or inside the casting near to the surface. Blowhole could be a pinhole which is very small or it could be a subsurface hole which can only be observed upon machining. Some of the reasons of forming blowholes are low permeability, improper venting, and higher moisture content of the sand mold.

Porosity: Porosity or microporosity is one of the major defects observed in most of the castings. It consists of small voids distributed throughout the casting. The air inside the mold cavity can be trapped during pouring. The air is further compressed and increase pressure as the pouring continues inside the cavity. When the mold is full, the entrapped air disperse as small spheres of high pressure which later elongates due to swirling flow.

Sand Burning: The chemical burn-on and metal penetration is referred to as sand burning or burn-on defect. It is usually observed in thick-walled castings where thin sand crust adheres to the casting surfaces. Primarily, this defect occurs due to sintering of bentonite and silicate components when exposed to high temperature melts. Also, formation of iron silicate reduces the sinter point of the sand. Sintering and melting accelerates the penetration of molten iron into the sand, the layers of which are then adhered firmly to the casting surface.

Misrun: Misrun is caused by incomplete filling of the mold by molten metal. This results in incomplete casting with smoothed or rounded edges formed on the castings. The causes of misruns are similar to that of cold shut defect.

Cold Shut: This defect is formed when two metal streams flow together but fails to fuse properly due to premature freezing. It is caused by low melting temperature of the melt or poor gating system in the mold.

Sand Inclusion: Inclusions are the defects that appear like a slag inside a metal casting. Sand or slag inclusion is also termed as scab or blacking scab [22]. These inclusions are formed near to the casting surfaces together with the metallic protuberances at other points. It is often difficult to associate these defects with a particular cause as they occur at widely varying positions in a casting. During pouring, the melt stream can tear some areas of sand which then floats to the surface of the casting as they cannot be wetted by the molten metal. These defects are commonly found in combination with blowholes and slag particles.

5.2 Shape Related Defects

Mismatch: The dislocation at parting line of the mold causes a mismatch defect. The shifting mold flashes causes misalignment in cope and drag portions of the mold due to which half of the final casting (either in cope or drag) is found to be slightly displaced over the plane of parting line.

Distortion or Warp: Shrinkage during solidification of castings introduces stresses inside the cast products. Distortion or warpage occurs due to non-uniform distribution of stresses throughout the casting. Warped parts are not visually or functionally acceptable for their desired applications.

Flash Defect: A flash defect is described as unwanted, excess metal which comes out of the die attached to the cavity or runner. It is appeared as a thin sheet of metal at the parting surfaces. This defect cannot be associated with a particular cause, however, the extent could classify the defect from a minor inconvenience to a severe quality concern.

5.3 Thermal Defects

Hot tears or cracks: Hot tearing or cracking occurs during the final stage of solidification or early stage of cooling when the casting is restrained from contraction by a permanent mold. The defect appears as separation of metal at a point of higher tensile stress caused by metal's inability to shrink naturally. This defects is minimized by collapsing the mold in sand casting and expendable mold processes, however, for permanent mold processes the casting must be removed immediately after solidification.

Shrinkage: This defect appears as a depression in the surface or an internal void in the casting which occurs when the feed metal is not available to compensate for shrinkage during solidification. There are two main types of shrinkage defects: open shrinkage and closed shrinkage. Open shrinkage defects are exposed to the atmosphere and take the shape of pipes or caved surfaces. Pipes forms at the surface and burrow into the castings whereas caved surfaces are the shallow cavities form across the surface of castings. Closed shrinkage refers to microporosity formed within the casting. Solidified metals contain isolated pools of liquid termed as hotspots. Shrinkage normally occurs at the top of the hotspots. Often the closed shrinkage is divided into macro- and micro-porosity depending upon the size of pores developed as a result of shrinkage.

5.4 Defects by Appearance

Defective Surface: This is a surface defect which develops in the form of flow marks as a result of flowing streams of molten metal within the mold. Sound castings with no internal flaw can also have this kind of surface defect.

Discontinuities: Cracks on the casting surfaces caused by rough handling or excessive temperature during shakeout operation.

Metallic Projections: This defect is similar to flash defect where flat projections of irregular thickness are formed perpendicular to one of the faces of castings. These projections are commonly observed along the joint or parting line of the mold, at a core print, or where two elements of the mold intersect.

Incomplete Castings: This defect is occurred due to shortage of molten metal during pouring. Normally, improper calculation of volume of the melt may result in this type of defect.

Cavities: Cavities in a cast product results from many defects already discussed above such as blowholes, pinholes, depression due to shrinkage etc. These cavities are not location specific and can be observed in all regions of cast products.

6   Design Considerations in Metal Casting

A good casting design often results in sound cast products free from defects. The casting design comprises of six steps [18] which are as follows.

1. Physical designing of the parts to be cast: size, shape, tolerances, dimensional changes during the process and others.
2. Material selection: mechanical and physical properties, castability, fluid flow characteristics, section size sensitivity.
3. Pattern making for mold and cores: gating and risers design, fluid flow and heat transfer.
4. Casting process selection: casting size, metal casting limitations, dimensional requirements, and production costs.
5. Post casting procedures: Machining, heat treatment etc.
6. Casting product evaluation and quality control.

Owing to the iterative nature of the process, it requires excellent communication between the personnel involved throughout the process. Figure 30 demonstrates a complete casting design process envelop where the needed communication is represented by the arrows. It is important for a casting designer to analyze the material for its properties and limitations, process capabilities and possible obstacles to produce that casting. The key considerations by a casting designer include but not limited to minimal changes in section size, minimization/elimination of sharp corners, understanding of mechanical properties desired in the cast product, tolerancing, locating and handling requirements, process limitations, machining requirements, and use of statistical methods for process control [18]. During casting design, the conventional way of making a part must be questioned and knowledge of new technologies should be employed to stretch the design envelop as much as possible.

The next phase is to decide upon if the part can be produced using the selected material and the casting process in an economical manner. Pattern makers and method engineers utilize the knowledge of fluid flow and heat transfer to design the mold and cores (if necessary) for the part to be cast. It is important at this stage to computationally analyze the filling and solidification sequence of the casting

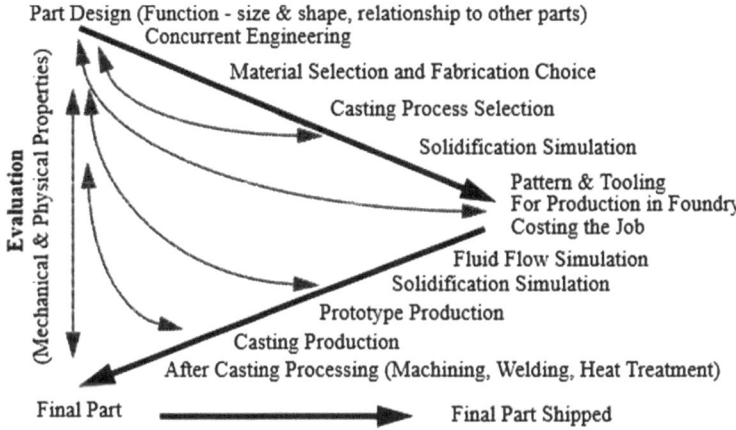

Fig. 30 Casting process design envelop [18]

process for which various casting simulation softwares are available. These simulations not only provide the insight of the process but also reduces the time between the design and prototype castings by optimizing the gating and riser system. This whole process requires designer, pattern engineer and the method engineer to be well communicated as shown by the dashed lines in Fig. 30. The final estimates of cost are provided by planning people in foundry after careful economic consideration. If the cost is within the realm of reality, it leads to the production of pattern for casting the final product. However, if the cost is too high, the design has to be modified critically from scratch. Some of the important considerations by a casting engineer during the process are evaluation of dimensional accuracy, quantification of microstructural integrity (presence of required microconstituents, casting defects etc.), understanding of response to machining, heat treatment or welding, determination of mechanical properties in critical sections etc.

In more recent years, rapid prototyping is being utilized to minimize production time for cast products. Fast pattern production is accomplished through technologies such as stereolithography, selective laser sintering, fused deposition modeling, laminated object manufacturing, direct shell production etc. Casting engineer must also aware of the dimensional tolerances and understand the effect of different casting processes on the dimensional accuracy of the cast products. Certainly, each process (green sand vs. lost foam vs. investment) provides a different attainable dimensional accuracy which must be understood prior to select the process for any particular product. It is evident from Fig. 30 the design improvements can be suggested at this stage to maximize production and minimize cost and difficulty of the process simultaneously.

The design considerations discussed above are documented in the form of industrial standards which are followed by foundrymen. Some of the well-known organizations for developing casting standards are American Society for Testing and Materials (ASTM), American Foundry Society (AFS), Society of Automotive

Engineers (SAE), and Japan International Cooperation Agency (JICA). These standards are developed for proper utilization of different cast materials and casting processes to produce castings used in various engineering applications such as valves, flanges, fittings, and other pressure containing parts for high- and low-temperature applications.

7 Conclusions

This book is an attempt to provide comprehensive knowledge of metalcasting technologies. The key conclusions derived from this work are as follows:

- Metalcasting has developed significantly over time as detailed in historical evolution section. The recent advances in metalcasting industry significantly rely on computational means to investigate the casting processes, their filling and solidification sequence, and life prediction in the presence of distortion and stresses in the cast parts.
- A range of casting processes has already developed to produce parts with high degree of intricacy, however, efforts are being made to maximize the casting yield and minimize the defects through continuous process development.
- Foundry practices such as mold materials and molding techniques, pattern making and cores, cast alloys, various type of furnaces, pouring, cleaning and finishing operations have to be well understood for streamlined production.
- Casting defects, their causes and remedies are important to understand by casting designer and engineers so that these defects can be reduced to minimum.
- The design considerations such as material properties, mechanical properties of cast products, pattern and core development, tolerancing, economic analysis, machining and cleaning requirements etc. need proper consideration for uninterruptable production of cast parts.

Acknowledgments This work is part of Ph.D. Dissertation under the supervision of Professor Anwar Khalil Sheikh at King Fahd University of Petroleum and Minerals (KFUPM). The authors greatly acknowledge the support provided by KFUPM in this research.

References

1. ASM Handbook Volume 15: Casting—ASM International.
2. AFS Technical Department. Timeline of casting technology.
3. Bak D. Rapid prototyping or rapid production? 3D printing processes move industry towards the latter. Assem Autom. 2003;23:340–5. doi:10.1108/01445150310501190.
4. Bawa HS. Manufacturing processes—II. New York: Tata McGraw-Hill Education; 2004.
5. Beeley PR. Foundry technology. Butterworth-Heinemann; 2001.

6. Butler WA, Timelli G, Battaglia E, Bonollo F. Die Casting (Permanent Mold). Ref. Module Mater. Sci. Mater. Eng., Elsevier; 2016.
7. Chhabra M, Singh R. Experimental investigation of pattern-less casting solution using additive manufacturing technique. MIT Int J Mech Eng. 2011;1:17–25.
8. Chirita G, Soares D, Silva FS. Advantages of the centrifugal casting technique for the production of structural components with Al–Si alloys. Mater Des. 2008;29:20–7. doi:10.1016/j.matdes.2006.12.011.
9. Ghomashchi MR, Vikhrov A. Squeeze casting: an overview. J Mater Process Technol. 2000;101:1–9. doi:10.1016/S0924-0136(99)00291-5.
10. Greer SE. A comparison of the ancient metal casting materials and processes to modern metal casting materials and processes. Hartford, Connecticut: Master of Mechanical Engineering. Rensselaer Polytechnic Institute; 2009.
11. Groover MP. Fundamentals of modern manufacturing: materials, processes, and systems. Hoboken: Wiley; 2010.
12. Jurrens KK. Standards for the rapid prototyping industry. Rapid Prototyp J. 1999;5:169–78. doi:10.1108/13552549910295514.
13. Kalpakjian S, Schmid S. Manufacturing engineering & technology. 7 ed. Upper Saddle River: Prentice Hall; 2013.
14. Kostakis V, Niaros V, Giotitsas C. Open source 3D printing as a means of learning: an educational experiment in two high schools in Greece. Telemat Inform. 2015;32:118–28. doi:10.1016/j.tele.2014.05.001.
15. Krouth TJ. Foundry tooling and metal castings in days. Proceedings of international conference worldwide advances in rapid and high-performance tooling, EuroMold, Frankfurt/M, Germany; 2002.
16. Li C, Wei Z, Chenhong W, Baochang J, Qingchun X. Application of digital pattern-less molding technology to produce art casting. Research and Development, China Foundry 2014;11. doi:672-6421(2014)06-487-06.
17. Mold and Core Test Handbook. American Foundrymen's Society; 1978.
18. Pattnaik S, Karunakar DB, Jha PK. Developments in investment casting process—a review. J Mater Process Technol. 2012;212:2332–48. doi:10.1016/j.jmatprotec.2012.06.003.
19. Rajkolhe R, Khan JG. Defects, causes and their remedies in casting process: a review. Int J Res Advent Technol. 2014;2:2321–963.
20. Ravi B. Metal casting: computer-aided design and analysis. PHI Learning Pvt. Ltd.; 2005.
21. Rundman KB. Metal casting. Dep Mater Sci Eng Mich Technol Univ Michagan 2000:17–9.
22. Schey JA. Introduction to manufacturing processes. New York: McGraw-Hill; 1987.
23. Simpson BL. History of the metal-casting industry. 2nd ed. Revised ed. American Foundrymen's Society; 1969.
24. Singh R, Singh JP. Comparison of rapid casting solutions for lead and brass alloys using three-dimensional printing. Proc Inst Mech Eng Part C J Mech Eng Sci. 2009;223:2117–23. doi:10.1243/09544062JMES1387.
25. Singh JP, Singh R. Investigations for a statistically controlled rapid casting solution of lead alloys using three-dimensional printing. Proc Inst Mech Eng Part C J Mech Eng Sci. 2009;223:2125–34. doi:10.1243/09544062JMES1337.
26. Swift KG, Booker JD. Chapter 3—Casting processes. In: Booker KGSD, editor. Manuf. Process Sel. Handb., Oxford: Butterworth-Heinemann; 2013, p. 61–91.
27. Wohlers T. Wohlers Report. State Ind Annu Worldw Prog Rep Wohlers Assoc Inc 2008:29–37.